4.4 The halogens and their compounds: Group 7   76
4.5 Extraction of metals   81
     Sample question and model answer   85
     Practice examination questions   86

Chapter 5   Chemical energetics

5.1 Enthalpy changes   87
5.2 Determination of enthalpy changes   90
5.3 Bond enthalpy   94
     Sample question and model answer   96
     Practice examination questions   97

Chapter 6   Rates and equilibrium

6.1 Reaction rates   98
6.2 Catalysis   102
6.3 Chemical equilibrium   104
6.4 Acids and bases   109
     Sample question and model answer   111
     Practice examination questions   112

Chapter 7   Organic chemistry

7.1 Basic concepts   113
7.2 Hydrocarbons from oil   118
7.3 Alkanes   120
7.4 Alkenes   122
7.5 Alcohols   127
7.6 Halogenoalkanes   132
     Sample question and model answer   135
     Practice examination questions   136

     Practice examination answers   138
     Periodic Table   143
     Index   144

Specification lists

AQA Chemistry

MODULE	SPECIFICATION TOPIC	CHAPTER REFERENCE	STUDIED IN CLASS	REVISED	PRACTICE QUESTIONS
Module 1 (M1) Atomic Structure, Bonding and Periodicity	Atomic structure	1.1, 1.2, 1.3			
	Amount of substance	2.1, 2.2, 2.3			
	Bonding	3.1, 3.2, 3.3, 3.4, 3.5			
	Periodicity	4.1, 4.3			
Module 2 (M2) Foundation Physical and Inorganic Chemistry	Energetics	5.1, 5.2, 5.3			
	Kinetics	6.1, 6.2			
	Equilibria	6.3			
	Redox reactions	4.2			
	Group VII: the halogens	4.4			
	Extraction of metals	4.5			
Module 3 (M3) Introduction to Organic Chemistry	Nomenclature and isomerism	7.1			
	Petroleum and alkanes	7.2, 7.3			
	Alkenes and epoxyethane	7.4			
	Haloalkanes	7.6			
	Alcohols	7.5			

Examination analysis

The specification comprises three unit tests. All questions are compulsory.

Unit 1	Structured questions: short and extended answers	1 hr test	30%
Unit 2	Structured questions: short and extended answers	1 hr test	30%
Unit 3	comprises two components:		
	3(a) Structured questions: short and extended answers	1 hr test	25%
	3(b) Centre-assessed coursework **or** practical examination	2 hr test	15%

AS

Chemistry

Rob Ritchie

Contents

Specification list   4

AS/A2 Level Chemistry courses   11

Different types of questions in AS examinations   13

Exam technique   15

What grade do you want?   17

Four steps to successful revision   18

Chapter 1 Atomic structure

1.1 Atoms and isotopes   19

1.2 The electronic structure of the atom   21

1.3 Experimental evidence for electronic structures   25

Sample question and model answer   28

Practice examination questions   29

Chapter 2 Atoms, moles and equations

2.1 Atomic mass   30

2.2 The mole   33

2.3 Formulae, equations and reacting quantities   37

Sample question and model answer   41

Practice examination questions   42

Chapter 3 Bonding and structure

3.1 Chemical bonding   44

3.2 Shapes of molecules   50

3.3 Electronegativity, polarity and polarisation   52

3.4 Intermolecular forces   55

3.5 Bonding structure and properties   58

Sample question and model answer   62

Practice examination questions   63

Chapter 4 The Periodic Table

4.1 The modern Periodic Table   64

4.2 Redox reactions   68

4.3 The s-block elements: Group 1 and Group 2   71

OCR Chemistry

MODULE	SPECIFICATION TOPIC	CHAPTER REFERENCE	STUDIED IN CLASS	REVISED	PRACTICE QUESTIONS
Module 2811 (M1) Foundation Chemistry	Atoms, molecules and stoichiometry	2.1, 2.2, 2.3			
	Atomic structure	1.1, 1.2, 1.3			
	Chemical bonding and structure	3.1, 3.2, 3.3, 3.4, 3.5			
	The Periodic Table: introduction	4.1			
	The Periodic Table: Group 2	4.2, 4.3			
	The Periodic Table: Group 7	4.4			
Module 2812 (M2) Chains and Rings	Basic concepts	7.1			
	Hydrocarbons: Alkanes	7.3			
	Hydrocarbons: Fuels	7.2			
	Hydrocarbons: Alkenes	7.4			
	Alcohols	7.5			
	Halogenoalkanes	7.6			
Module 2813/1 (M3) How Far, How Fast?	Enthalpy changes	5.1, 5.2, 5.3			
	Reaction rates	6.1, 6.2			
	Chemical equilibrium	6.3, 6.4			

Examination analysis

The specification comprises three unit tests. All questions are compulsory.

Unit 2811 Structured questions: short and extended answers — 1 hr test — 30%

Unit 2812 Structured questions: short and extended answers — 1 hr test — 30%

Unit 2813 comprises two components:
- 2813/1: Structured questions: short and extended answers — 45 min test — 20%
- 2813/2: Centre-assessed coursework **or**
- 2813/3: Practical examination — 1 hr 30 min test — 20%

OCR Chemistry, Salters

MODULE	SPECIFICATION TOPIC	CHAPTER REFERENCE	STUDIED IN CLASS	REVISED	PRACTICE QUESTIONS
Module 2850 (M1) Chemistry for life	The elements of life	1.1, 1.2, 1.3, 2.1, 2.2, 2.3, 3.1, 3.2, 4.1, 4.3			
	Developing fuels	2.2, 2.3, 5.1, 5.2, 5.3, 6.2, 7.1, 7.2, 7.3			
Module 2848 (M2) Chemistry of Natural Resources	Minerals to elements	1.2, 2.3, 3.5, 4.2, 4.4, 6.4			
	The atmosphere	6.1, 6.2, 6.3, 7.1, 7.3, 7.6			
	The polymer revolution	3.3, 3.4, 7.1, 7.4, 7.5			

Note that the Salters approach 'drip-feeds' concepts throughout the course. For example, you will study different aspects of Chapter 3: Bonding and structure in different topics. You may need to revisit the chapters and practice questions in this study guide throughout your AS course.

Examination analysis

The specification comprises three unit tests. All questions are compulsory.

Unit 2850 Structured questions: short and extended answers   *1 hr 15 min test*   *30%*

Unit 2848 Structured questions: short and extended answers   *1 hr 30 min test*   *40%*

Unit 2853 comprises two components:

- Open-book paper   Externally assessed coursework   *15%*
- Experimental skills   Internally assessed coursework   *15%*

The open-book paper is taken over a two week period during the third term of the AS course at a date specified by OCR. Marks for this paper take into account quality of written communication.

Edexcel Chemistry

MODULE	SPECIFICATION TOPIC	CHAPTER REFERENCE	STUDIED IN CLASS	REVISED	PRACTICE QUESTIONS
Module 1 (M1) *Structure, bonding and main group chemistry*	Atomic structure	1.1, 1.2, 1.3			
	Formulae, equations and moles	2.1, 2.2, 2.3			
	Structure and Bonding	3.1, 3.2, 3.3, 3.4, 3.5			
	The Periodic Table Introduction	4.1			
	Introduction to reduction and oxidation	4.2			
	Group 1 and Group 2	4.3			
	Group 7	4.4			
Module 2 (M2) *Introductory organic chemistry, energetics, kinetics and equilibrium and applications*	Energetics I	5.1, 5.2, 5.3			
	Organic chemistry (Introduction)	7.1, 7.3, 7.4, 7.5, 7.6			
	Kinetics I	6.1, 6.2			
	Chemical Equilibria I	6.3			
	Industrial inorganic chemistry	4.4, 4.5, 6.3			

Examination analysis

The specification comprises three unit tests. All questions are compulsory.

Unit 1	Structured questions: short answers	*1 hr test*	30%
Unit 2	Structured questions: short and extended answers	*1 hr test*	30%
Unit 3	Laboratory skills comprises two components:		
3A	Centre-assessed coursework **or** practical examintion	*1 hr 45 min test*	20%
3B	Structured questions	*1hr test*	20%

    Unit 3B assesses a students' ability to interpret information generated and drawn from experimental situations in the laboratory.

Edexcel Chemistry (Nuffield)

MODULE	SPECIFICATION TOPIC	CHAPTER REFERENCE	STUDIED IN CLASS	REVISED	PRACTICE QUESTIONS
Module 1 (M1) Introductory chemistry	Metal compounds	1.1, 2.1, 2.2, 2.3, 5.1			
	Organic chemistry: alcohols	7.1, 7.5			
	The Periodic Table and atomic structure	1.2, 1.3, 3.1, 3.5, 4.1			
	Acid-base reactions and Group 2	2.2, 2.3, 4.1, 4.3, 6.4			
	Energy changes and reactions	5.1, 5.2			
Module 2 (M2) Bonding and reactions	Redox reactions and the halogens	4.2, 4.4			
	Covalency and bond breaking	3.1, 3.2, 3.3, 5.1, 5.2, 5.3, 6.1, 6.2, 6.3			
	Organic chemistry: hydrocarbons	7.1, 7.2, 7.3, 7.4			
	Intermolecular forces	3.4, 3.5			
	Organic chemistry: halogenoalkanes	7.6			

Examination analysis

The specification comprises three unit tests. All questions are compulsory.

Unit 1	Structured questions: short and extended answers	1 hr 15 min test	30%
Unit 2	Structured questions and comprehension exercise	1 hr 30 min test	40%

Section 2A will contain structured questions, based on topics 6 to 10 with a prerequisite of understanding of topics 1 to 5.

Section 2B will consist of a summary/comprehension exercise based on a passage. This exercise will include marks for quality of written communication.

Unit 3	Centre-assessed coursework		30%

WJEC Chemistry

MODULE	SPECIFICATION TOPIC	CHAPTER REFERENCE	STUDIED IN CLASS	REVISED	PRACTICE QUESTIONS
Module CH1 (CH1) Physical – Inorganic	Atomic structure	1.1, 1.2, 1.3			
	The mole and stoichiometry	2.1, 2.2, 2.3			
	Structure and bonding	3.1, 3.2, 3.3, 3.4			
	Gases, liquids and solids	3.5			
	The Periodic Table	4.1, 4.2, 4.3, 4.4			
Module CH2 (CH2) Physical – Organic	Principles of energetics	5.1, 5.2, 5.3			
	Organic compounds	7.1, 7.5, 7.6			
	Hydrocarbons and petroleum	7.2, 7.3, 7.4			
	Chemical and acid-base equilibria	2.3, 6.3, 6.4			
	Chemical kinetics	6.1, 6.2			
	Industrial and environmental aspects	6.2, 6.3			

Examination analysis

The specification comprises three unit tests. All questions are compulsory.

Unit CH1	Structured questions and objective questions	1 hr 30 min test	35%
Unit CH2	Structured questions and objective questions	1 hr 30 min test	35%
Unit CH3	Experimental skills comprising two components:		
	CH3a Structured questions: short answers	1 hour test	10%
	CH3b Centre-assessed coursework		20%

Unit CH3a assesses the link between theory and experiment, using the content of teaching modules CH1 and CH2 and other related experiments.

NICCEA Chemistry

MODULE	SPECIFICATION TOPIC	CHAPTER REFERENCE	STUDIED IN CLASS	REVISED	PRACTICE QUESTIONS
Module M1 (M1) General Chemistry	Atomic structure	1.1, 1.2, 1.3, 2.1			
	Bonding	3.1, 3.3, 3.5			
	Shapes of molecules	3.2			
	Intermolecular forces	3.4, 3.5			
	Energetics	5.1, 5.2, 5.3			
	Redox	4.2			
	Periodic Table	4.1			
	Group 7	4.4			
	Titrations	2.3, 4.4			
Module M2 (M2) Organic, Physical and Inorganic Chemistry	Isomerism	7.1			
	Hydrocarbons	7.2			
	Alkanes	7.3			
	Alkenes	7.4			
	Halogenoalkanes	7.6			
	Alcohols	7.5			
	Equilibrium	6.3			
	Kinetics	6.1, 6.2			
	Group 2	4.3			

Note that the use of balanced equations and calculations involving the mole concept (Topic refs: 2.1, 2.2 and 2.3) are general aspects that apply to all the molecules, particularly Module M1.

Examination analysis

The specification comprises three unit tests. All questions are compulsory.

Unit 1	Multiple-choice questions and structured questions	*1 hr 30 min test*	*35%*
Unit 2	Multiple-choice questions and structured questions	*1 hr 30 min test*	*35%*
Unit 3	Practical examination	*2 hr 30 min test*	*30%*

AS/A2 Level Chemistry courses

AS and A2

All Chemistry A Level courses being studied from September 2000 are in two parts, with three separate modules in each part. Students first study the AS (Advanced Subsidiary) course. Some will then go on to study the second part of the A Level course, called A2. Advanced Subsidiary is assessed at the standard expected halfway through an A Level course: i.e., between GCSE and Advanced GCE. This means that new AS and A2 courses are designed so that difficulty steadily increases:

- AS Chemistry builds from GCSE Science
- A2 Chemistry builds from AS Chemistry.

How will you be tested?

Assessment units

For AS Chemistry, you will be tested by three assessment units. For the full A Level in Chemistry, you will take a further three units. AS Chemistry forms 50% of the assessment weighting for the full A Level.

Each unit can normally be taken in either January or June. Alternatively, you can study the whole course before taking any of the unit tests. There is a lot of flexibility about when exams can be taken and the diagram below shows just some of the ways that the assessment units may be taken for AS and A Level Chemistry.

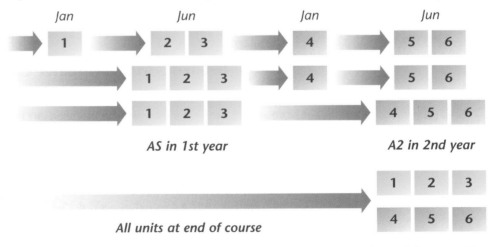

If you are disappointed with a module result, you can resit each module once. You will need to be very careful about when you take up a resit opportunity because you will have only one chance to improve your mark. The higher mark counts.

A2 and Synoptic assessment

After having studied AS Chemistry, you may wish to continue studying Chemistry to A Level. For this you will need to take three further units of Chemistry at A2. Similar assessment arrangements apply except some units, those that draw together different parts of the course in a synoptic assessment, have to be assessed at the end of the course.

Coursework

Coursework may form part of your A Level Chemistry course, depending on which specification you study. Where students have to undertake coursework, it is usually for the assessment of practical skills but this is not always the case. See pages 4–10.

Key Skills

To gain the key skills qualification, which is equivalent to an AS Level, you will need to collect evidence together in a 'portfolio' to show that you have attained a satisfactory level in Communication, Application of number and Information technology. You will also need to take a formal testing in each key skill. You will have many opportunities during AS Chemistry to develop your key skills.

What skills will I need?

The assessment objectives for AS Chemistry are shown below.

Knowledge with understanding

- recall of facts, terminology and relationships
- understanding of principles and concepts
- drawing on existing knowledge to show understanding of the responsible use of chemistry in society
- selecting, organising and presenting information clearly and logically

Application of knowledge and understanding, analysis and evaluation

- explaining and interpreting principles and concepts
- interpreting and translating, from one form into another, data presented as continuous prose or in tables, diagrams and graphs
- carrying out relevant calculations
- applying knowledge and understanding to familiar and unfamiliar situations
- assessing the validity of chemical information, experiments, inferences and statements

Experimental and investigative skills

Chemistry is a practical subject and part of the assessment of AS Chemistry will test your practical skills. You will be assessed on four main skills:

- planning
- implementing
- analysing evidence and drawing conclusions
- evaluating evidence and procedures.

The skills may be assessed in the context of separate practical exercises, although more than one skill could be assessed in any one exercise. They may also be assessed all together in the context of a single 'whole investigation'. An investigation may be set by your teacher or you may be able to pursue an investigation of your choice. Alternatively, you may take a practical examination

You will receive guidance about how your practical skills will be assessed from your teacher. This study guide concentrates on preparing you for the written examinations testing the subject content of AS Chemistry.

Different types of questions in AS examinations

In AS Chemistry examinations, different types of question are used to assess your abilities and skills. Unit tests mainly use structured questions requiring both short answers and more extended answers.

Short-answer questions

A short-answer question may test recall or it may test understanding. Short-answer questions normally have space for the answers printed on the question paper.

Here are some examples (the answers are shown in *italics*):

What is meant by isotopes?

Atoms of the same element with different masses.

Calculate the amount (in mol) of H_2O in 4.5 g of H_2O.

1 mol H_2O has a mass of 18 g. ∴ 4.5 g of H_2O contains 4.5/18 = 0.25 mol H_2O.

Structured questions

Structured questions are in several parts. The parts usually have a common context and they often become progressively more difficult and more demanding as you work your way through the question. A structured question may start with simple recall, then test understanding of a familiar or an unfamiliar situation.

Most of the practice questions in this book are structured questions, as this is the main type of question used in the assessment of AS Chemistry.

When answering structured questions, do not feel that you have to complete one question before starting the next. The further you are into a question, the more difficult the marks are to obtain. If you run out of ideas, go on to the next question. You need to respond to as many parts of questions on an exam paper as possible. You will not score well if you spend so long trying to perfect the first questions that you do not have time to attempt later questions.

Here is an example of a structured question that becomes progressively more demanding.

(a) Write down the atomic structure of the two isotopes of potassium: ^{39}K and ^{41}K.

    (i) ^{39}K ...19... protons; ...20... neutrons; ...19... electrons. ✓

    (ii) ^{41}K ...19... protons; ...22... neutrons; ...19... electrons. ✓   [2]

(b) A sample of potassium has the following percentage composition by mass: ^{39}K: 92%; ^{41}K: 8% Calculate the relative atomic mass of the potassium sample.

    92 × 39/100 + 8 × 41/100 = 39.16 ✓   [1]

(c) What is the electronic configuration of a potassium atom?

    $1s^2 2s^2 2p^6 3s^2 3p^6 4s^1$ ✓   [1]

(d) The second ionisation energy of potassium is much larger than its first ionisation energy.

(i) Explain what is meant by the *first ionisation energy* of potassium.

The energy required to remove an electron ✓ *from each atom in 1 mole* ✓ *of gaseous atoms* ✓

(ii) Why is there a large difference between the values for the first and the second ionisation energies of potassium?

The 2nd electron removed is from a different shell ✓ *which is closer to the nucleus and experiences more attraction from the nucleus.* ✓ *This outermost electron experiences less shielding from the nucleus because there are fewer inner electron shells than for the 1st ionisation energy.* ✓ [6]

Extended answers

In AS Chemistry, questions requiring more extended answers may form part of structured questions or may form separate questions. They may appear anywhere on the paper and will typically have between 5 and 10 marks allocated to the answers as well as several lines of answer space. These questions are also often used to assess your abilities to communicate ideas and put together a logical argument.

The correct answers to extended questions are often less well-defined than to those requiring short answers. Examiners may have a list of points for which credit is awarded up to the maximum for the question.

An example of a question requiring an extended answer is shown below.

Magnesium oxide and sulphur trioxide are both solids at 10 °C. Magnesium oxide has a melting point of 2852 °C. Sulphur trioxide has a melting point of 17 °C. Explain, in terms of structure and bonding, why these two compounds have such different melting points. [7]

Points that the examiners might look for include:

Magnesium oxide has a giant lattice structure ✓ *containing ionic bonds* ✓ *These strong forces need to be broken during melting to allow ions to move free of the rigid lattice.* ✓ *This requires a large amount of energy supplied by a high temperature.* ✓

Sulphur trioxide has a simple molecular structure ✓ *held together by van der Waals' forces.* ✓ *These weak forces need to be broken during melting to allow molecules to move free of the rigid lattice.* ✓ *This requires a small amount of energy supplied by a low temperature.* ✓

8 marking points → [7]

In this type of response, there may be an additional mark for a clear, well-organised answer, using specialist terms. In addition, marks may be allocated for legible text with accurate spelling, punctuation and grammar.

Other types of questions

Free-response and open-ended questions allow you to choose the context and to develop your own ideas. These are little used for assessing AS Chemistry but, if you decide to take the full A Level in Chemistry, you will encounter this type of question during synoptic assessment.

Multiple-choice or objective questions require you select the correct response to the question from a number of given alternatives. These are rarely used for assessing AS Chemistry, although it is possible that some short-answer questions will use a multiple-choice format.

Exam technique

Links from GCSE

Advanced Subsidiary Chemistry builds from grade CC in GCSE Science: Double Award, or equivalent in Science: Chemistry. This study guide has been written so that you will be able to tackle AS Chemistry from a GCSE Science background.

You should not need to search for important Chemistry from GCSE Science because this has been included where needed in each chapter. If you have not studied Science for some time, you should still be able to learn AS Chemistry using this text alone.

What are examiners looking for?

Examiners use instructions to help you to decide the length and depth of your answer.

If a question does not seem to make sense, you may have misread it – read it again!

State, define or list

This requires a short, concise answer, often recall of material that can be learnt by rote.

Explain, describe or discuss

Some reasoning or some reference to theory is required, depending on the context.

Outline

This implies a short response, almost a list of sentences or bullet points.

Predict or deduce

You are not expected to answer by recall but by making a connection between pieces of information.

Suggest

You are expected to apply your general knowledge to a 'novel' situation, one which you have not directly studied during the AS Chemistry course.

Calculate

This is used when a numerical answer is required. You should always use units in quantities and significant figures should be used with care.

Look to see how many significant figures have been used for quantities and give your answer to this degree of accuracy.

If the question uses three significant figures, then give your answer in three significant figures also.

Some dos and don'ts

Dos

Do answer the question

No credit can be given for good Chemistry that is irrelevant to the question.

Do use the mark allocation to guide how much you write

Two marks are awarded for two valid points – writing more will rarely gain more credit and could mean wasted time or even contradicting earlier valid points.

Do use diagrams, equations and tables in your responses

Even in 'essay-type' questions, these offer an excellent way of communicating chemistry.

Do write legibly

An examiner cannot give marks if the answer cannot be read.

Do write using correct spelling and grammar. Structure longer essays carefully

Marks are now awarded for the quality of your language in exams.

Don'ts

Don't fill up any blank space on a paper

In structured questions, the number of dotted lines should guide the length of your answer.

If you write too much, you waste time and may not finish the exam paper. You also risk contradicting yourself.

Don't write out the question again

This wastes time. The marks are for the answer!

Don't contradict yourself

The examiner cannot be expected to choose which answer is intended. You could lose a hard-earned mark, e.g.

A covalent bond is a shared pair of electrons ✓ bonded by the electrostatic attraction between ions. ✗

Don't spend too much time on a part that you find difficult

You may not have enough time to complete the exam. You can always return to a difficult part if you have time at the end of the exam.

What grade do you want?

Everyone would like to improve their grades but you will only manage this with a lot of hard work and determination. You should have a fair idea of your natural ability and likely grade in Chemistry and the hints below offer advice on improving that grade.

For a Grade A

You will need to be a very good all-rounder.

- You must go into every exam knowing the work extremely well.
- You must be able to apply your knowledge to new, unfamiliar situations.
- You need to have practised many, many exam questions so that you are ready for the type of question that will appear.

The exams test all areas of the syllabus and any weaknesses in your Chemistry will be found out. There must be no holes in your knowledge and understanding. For a Grade A, you must be competent in all areas.

For a Grade C

You must have a reasonable grasp of Chemistry but you may have weaknesses in several areas and you may be unsure of some of the reasons for the Chemistry.

- Many Grade C candidates are just as good at answering questions as the Grade A students but holes and weaknesses often show up in just some topics.
- To improve, you will need to master your weaknesses and you must prepare thoroughly for the exam. You must become a better all-rounder.

For a Grade E

You cannot afford to miss the easy marks. Even if you find Chemistry difficult to understand and would be happy with a Grade E, there are plenty of questions in which you can gain marks.

- You must memorise all definitions.
- you must practise exam questions to give yourself confidence that you do know some Chemistry. In exams, answer the parts of questions that you know first. You must not waste time on the difficult parts. You can always go back to these later.
- the areas of Chemistry that you find most difficult are going to be hard to score on in exams. Even in the difficult questions, there are still marks to be gained. Show your working in calculations because credit is given for a sound method. You can always gain some marks if you get part of the way towards the solution.

What marks do you need?

The table below shows how your average mark is transferred into a grade.

average	80%	70%	60%	50%	40%
grade	A	B	C	D	E

Four steps to successful revision

Step 1: Understand

- Study the topic to be learned slowly. Make sure you understand the logic or important concepts.
- Mark up the text if necessary – underline, highlight and make notes.
- Re-read each paragraph slowly.

GO TO STEP 2

Step 2: Summarise

- Now make your own revision note summary:
  What is the main idea, theme or concept to be learned?
  What are the main points? How does the logic develop?
  Ask questions: Why? How? What next?
- Use bullet points, mind maps, patterned notes.
- Link ideas with mnemonics, mind maps, crazy stories.
- Note the title and date of the revision notes
  (e.g. Chemistry: Atomic structure, 3rd March).
- Organise your notes carefully and keep them in a file.

This is now in **short-term memory**. You will forget 80% of it if you do not go to Step 3.
GO TO STEP 3, but first take a 10 minute break.

Step 3: Memorise

- Take 25 minute learning 'bites' with 5 minute breaks.
- After each 5 minute break test yourself:
  Cover the original revision note summary
  Write down the main points
  Speak out loud (record on tape)
  Tell someone else
  Repeat many times.

The material is well on its way to **long-term memory**.
You will forget 40% if you do not do step 4. **GO TO STEP 4**

Step 4: Track/Review

- Create a Revision Diary (one A4 page per day).
- Make a revision plan for the topic, e.g. 1 day later, 1 week later, 1 month later.
- Record your revision in your Revision Diary, e.g.
  Chemistry: Atomic Structure, 3rd March 25 minutes
  Chemistry: Atomic Structure, 5th March 15 minutes
  Chemistry: Atomic Structure, 3rd April 15 minutes
  ... and then at monthly intervals.

Chapter 1
Atomic structure

The following topics are covered in this chapter:

- Atoms and isotopes
- The electronic structure of the atom
- Experimental evidence for electronic structures

1.1 Atoms and isotopes

After studying this section you should be able to:

- recall the relative charge and mass of a proton, neutron and electron
- explain the existence of isotopes
- use atomic number and mass number to determine atomic structure

LEARNING SUMMARY

What is an atom?

AQA	M1	OCR	M1
EDEXCEL	M1	SALTERS	M1
NICCEA	M1	WJEC	CH1
NUFFIELD	M1		

An atom is the smallest part of an element that can exist on its own. Atoms are so tiny that there are more atoms in a full stop than there are people in the world. It is now possible to see individual atoms by using the most modern and powerful microscopes. However, nobody has yet seen *inside* an atom and we must devise models for the structure of an atom from experimental evidence.

electron shells

nucleus (protons and neutrons)

Sub-atomic particles

We use a model in which an atom is composed of three sub-atomic particles: protons, neutrons and electrons.

In this model, protons and neutrons form the nucleus at the centre of the atom with electrons orbiting in shells. Compared with the total volume of an atom, the nucleus is tiny and extremely dense. Most of an atom is empty, made up of the space between the nucleus and the electron shells.

All matter is made up from the chemical elements and 115 are now known (although others will inevitably be discovered). The number of protons in the nucleus distinguishes the atoms of each element.

> An important principle of chemistry is the link between an element and the number of protons in its atoms.

KEY POINT

Each atom of an element has the same number of protons.
The number of protons in an atom of an element is called the atomic number, Z.

Properties of protons, neutrons and electrons

Some properties of protons, neutrons and electrons are shown in the table below.

> All atoms of hydrogen contain 1 proton.
>
> All atoms of carbon contain 6 protons.
>
> All atoms of oxygen contain 8 protons.

particle	relative mass	relative charge
proton, p	1	1+
neutron, n	1	0
electron, e	1/1840	1–

A proton has virtually the same mass as a neutron and, for most of chemistry, their masses can be assumed to be identical.

An electron has negligible mass compared with the mass of a proton or a neutron. In most of chemistry, the mass of an electron can be ignored.

The charge on a proton is opposite to that of an electron but each charge has the same magnitude. An atom is electrically neutral. To balance out the charges, an atom must have the same number of protons as electrons.

> An atom is electrically neutral.
> An atom contains the same number of protons as electrons.

All atoms of hydrogen contain:
1 proton and 1 electron.

All atoms of carbon contain:
6 protons and 6 electrons.

All atoms of oxygen contain:
8 protons and 8 electrons.

Isotopes

Without neutrons, the nucleus would just contain positively-charged protons. Like-charges repel, and a nucleus containing just protons would fly apart! Strong nuclear forces act between protons and neutrons and these hold the nucleus together. Neutrons can be thought of as 'nuclear glue'.

In each atom of an element, the number of protons in the nucleus is fixed. However, most elements contain atoms with different numbers of neutrons. These atoms are called isotopes and, because they have different numbers of neutrons, they also have different masses.

> Isotopes are atoms of the same element with different masses.
> Isotopes of an element have the same number of protons and electrons.
> Each isotope has a different number of neutrons in the nucleus.
> The combined number of protons and neutrons in an isotope of an element is called the mass number, A.

KEY POINT

The isotopes of an element react in the same way. This is because chemical reactions involve electrons – neutrons make no difference.

Chemists represent the nucleus of an isotope in a special way.

mass number
(protons + neutrons)

$$_Z^A X$$

atomic number
(protons)

The isotopes of carbon

Carbon exists as three isotopes, $_6^{12}C$, $_6^{13}C$, $_6^{14}C$. In all but the most accurate work, it is reasonable to assume that the relative mass of an isotope is equal to its mass number. It is easy to work out the atomic structure of an isotope from the representation above.

isotope	atomic number	mass number	protons	neutrons	electrons
$_6^{12}C$	6	12	6	6	6
$_6^{13}C$	6	13	6	7	6
$_6^{14}C$	6	14	6	8	6

Because all carbon isotopes have 6 protons, it is common practice to omit the atomic number, and $_6^{12}C$ is often shown as ^{12}C, or even as carbon-12.

Progress check

How many protons, neutrons and electrons are in the following isotopes?

1 $_3^7Li$; 2 $_{11}^{23}Na$; 3 $_9^{19}F$; 4 $_{13}^{27}Al$; 5 $_{26}^{55}Fe$.

1 $_3^7Li$: 3p, 4n, 3e
2 $_{11}^{23}Na$: 11p, 12n, 11e
3 $_9^{19}F$: 9p, 10n, 9e;
4 $_{13}^{27}Al$: 13p, 14n, 13e;
5 $_{26}^{55}Fe$: 26p, 29n, 26e.

1.2 The electronic structure of the atom

After studying this section you should be able to:

- *understand the relationship between shells, sub-shells and orbitals*
- *understand how atomic orbitals are filled*
- *describe electronic structure in terms of shells and sub-shells*

Energy levels or 'shells'

AQA	M1	OCR	M1
EDEXCEL	M1	SALTERS	M1
NICCEA	M1	WJEC	CH1
NUFFIELD	M1		

energy

$n = 4$ —— 32e⁻ ——

$n = 3$ —— 18e⁻ ——

$n = 2$ —— 8e⁻ ——

$n = 1$ —— 2e⁻ ——

You can imagine a model of an atom with electrons orbiting in shells around the nucleus. The electrons in each successive shell have an orbit further away from the nucleus. The further a shell is from the nucleus, the greater the shell's energy level. Each energy level is given a number called the principal quantum number, n. The shell closest to the nucleus has an energy level with $n = 1$, then $n = 2$ and so on.

The number of electrons that can occupy the first four energy levels is shown below:

n	shell	electrons
1	1st shell	2
2	2nd shell	8
3	3rd shell	18
4	4th shell	32

> The principal quantum shell, n, is the shell number.

If you look closely, you can see a pattern. The number of electrons that can occupy a shell is $2n^2$.

Using this model, electrons can be placed into available shells, starting with the lowest energy level. Each shell must be full before the next starts to fill. The table below shows how the shells are filled for the first 11 elements in the Periodic Table.

element	atomic number	electrons		
		$n = 1$	$n = 2$	$n = 3$
H	1	1		
He	2	2		
Li	3	2	1	
Be	4	2	2	
B	5	2	3	
C	6	2	4	
N	7	2	5	
O	8	2	6	
F	9	2	7	
Ne	10	2	8	
Na	11	2	8	1

Electronic energy levels are like a ladder and are filled from the bottom up.

Notice that the gap between successive energy levels becomes less with increasing energy.

This model breaks down as the $n = 3$ energy level is filled because each shell consists of sub-shells.

Sub-shells and orbitals

AQA	M1	OCR	M1
EDEXCEL	M1	SALTERS	M2
NICCEA	M1	WJEC	CH1
NUFFIELD	M1		

A more advanced model of electron structure is used in which each shell is made up of **sub-shells**.

Sub-shells

There are different types of sub-shell: s, p, d and f. Each type of sub-shell can hold a different number of electrons:

sub-shell	electrons
s	2
p	6
d	10
f	14

The table below shows the shells and sub-shells for the first four principal quantum numbers.

> Notice the labelling. The s sub-shell in the 2nd shell is lábelled 2s.

n	shell	sub-shell				total number of electrons	
1	1st shell	1s				2	= 2
2	2nd shell	2s	2p			2 + 6	= 8
3	3rd shell	3s	3p	3d		2 + 6 + 10	= 18
4	4th shell	4s	4p	4d	4f	2 + 6 + 10 + 14	= 32

- Each successive shell contains a new type of sub-shell.
- The 1st shell contains 1 sub-shell, the second sub-shell contains 2 sub-shells, and so on.

Orbitals

How do the electrons fit into the sub-shells? Mathematicians have worked out that electrons occupy negative charge clouds called orbitals and these make up each sub-shell.

> **KEY POINT**
> - An orbital can hold up to two electrons.
> - Each type of sub-shell has different orbitals: s, p, d and f.

The table below shows how electrons fill the orbitals in each sub-shell.

sub-shell	orbitals	electrons
s	1	1 x 2 = 2
p	3	3 x 2 = 6
d	5	5 x 2 = 10
f	7	7 x 2 = 14

s-orbitals

> One s orbital

- An s-orbital has a spherical shape.

p-orbitals

> Three p orbitals

- A p-orbital has a 3-dimensional dumb-bell shape.
- There are three p-orbitals, p_x, p_y and p_z, at right angles to one another.

Three p-orbitals

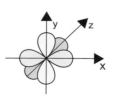

giving

d-orbitals and f-orbitals

The structures of d and f-orbitals are more complex.

Five d orbitals
Seven f orbitals

- There are five d-orbitals.
- There are seven f-orbitals.

How do two electrons fit into an orbital?

'Electrons in a box'

allowed

↑↓ ↓↑

not allowed

↑↑ ↓↓

Electrons are negatively charged and so they repel one another. An electron also has a property called **spin**. The two electrons in an orbital have opposite spins, helping to counteract the natural repulsion between their negative charges.

> - Chemists often represent an orbital as a box that can hold up to 2 electrons.
> - Each electron is shown as an arrow, indicating its spin: either ↑ or ↓.
> - Within an orbital, the electrons must have opposite spins.

KEY POINT

Filling the sub-shells

AQA	M1	OCR	M1
EDEXCEL	M1	SALTERS	M2
NICCEA	M1	WJEC	CH1
NUFFIELD	M1		

Within a shell, the sub-shell energies are in the order: s, p, d and f.

Note that the 4s sub-shell is at a lower energy than the 3d sub-shell.

The 4s sub-shell fills before the 3d sub-shell.

Sub-shells have different energy levels. The diagram below shows the relative energies for the sub-shells in the first four shells.

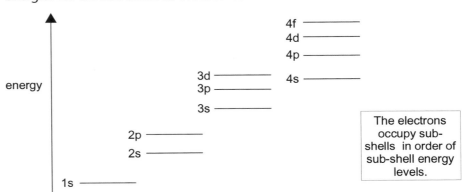

The electrons occupy sub-shells in order of sub-shell energy levels.

> - Shells and sub-shells are occupied in energy-level order.
> - Electrons occupy orbitals singly to prevent any repulsion caused by pairing.

KEY POINT

The 2s orbital is occupied before the 2p orbitals because it is at a lower energy.

Note that the 2p orbitals are occupied singly before pairing begins.

The diagram below shows how electrons occupy orbitals from boron to oxygen.

Use these 4 examples to see how electronic configuration is written.

The electronic configuration of an atom is a shorthand method showing how electrons occupy sub-shells.

The diagram below shows how the sub-shells are filled in an atom of potassium. Notice why the 4s sub-shell starts to fill before the 3d sub-shell.

The 19 electrons of K

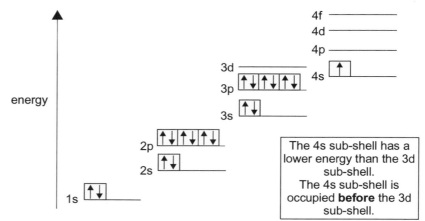

The 4s sub-shell has a lower energy than the 3d sub-shell.
The 4s sub-shell is occupied **before** the 3d sub-shell.

Sub-shells and the Periodic Table

AQA	M1	OCR	M1
EDEXCEL	M1	SALTERS	M2
NICCEA	M1	WJEC	CH1
NUFFIELD	M1		

You should be able to work out the electronic configuration for any element in the first four periods, i.e. up to krypton ($Z = 36$).

The Periodic Table is structured in blocks of 2, 6, 10 and 14, linked to sub-shells. The order of sub-shell filling can be seen by dividing the Periodic Table into blocks.

	1s		1s
s-block			**p-block**
2s			2p
3s		**d-block**	3p
4s		3d	4p
5s		4d	5p
6s		5d	6p
7s			

f-block
4f
5f

Simplifying electronic configurations

The similar electronic configurations within a group of the Periodic Table can be emphasised with a simpler representation in terms of the previous noble gas.

The electronic configurations for the elements in Group 1 are shown below:

Li: $1s^2 2s^1$ or $[He]2s^1$
Na: $1s^2 2s^2 2p^6 3s^1$ or $[Ne]3s^1$
K: $1s^2 2s^2 2p^6 3s^2 3p^6 4s^1$ or $[Ar]4s^1$

Progress check

1 Write the full electronic configuration in terms of sub-shells for:
  (a) C;   (b) Al;   (c) Ca;   (d) Fe;   (e) Br.

1 (a) C: $1s^2 2s^2 2p^2$
  (b) Al: $1s^2 2s^2 2p^6 3s^2 3p^1$
  (c) Ca: $1s^2 2s^2 2p^6 3s^2 3p^6 4s^2$
  (d) Fe: $1s^2 2s^2 2p^6 3s^2 3p^6 3d^6 4s^2$
  (e) Br: $1s^2 2s^2 2p^6 3s^2 3p^6 3d^{10} 4s^2 4p^5$

1.3 Experimental evidence for electronic structures

After studying this section you should be able to:

- *recall the definitions for first and successive ionisation energies*
- *understand the factors affecting the sizes of ionisation energies*
- *show how ionisation energies provide evidence for shells and sub-shells*

Ions

Positive ions form when electrons are lost.

Negative ions form when electrons are gained.

An atom can either lose or gain electrons to form an ion: a charged atom.

A positive ion is formed when an atom loses electrons. For example, a lithium atom forms a positive ion with a 1+ charge by losing an electron:

$$Li \longrightarrow Li^+ + e^-$$
$$(3p^+, 4n, 3e^-) \quad (3p^+, 4n, 2e^-)$$

A negative ion is formed when an atom gains electrons. For example, an oxygen atom forms a negative ion with a 2– charge by gaining two electrons:

$$O + 2e^- \longrightarrow O^{2-}$$
$$(6p^+, 6n, 6e^-) \qquad\qquad (6p^+, 6n, 8e^-)$$

Ionisation energy

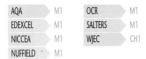

Ionisation energy measures the ease with which electrons are lost in the formation of positive ions. An element has as many ionisation energies as there are electrons.

> **KEY POINT**
>
> The **first** *ionisation energy* of an element is the energy required to remove **1 electron** from each atom in **1 mole** of **gaseous atoms** to form 1 mole of gaseous 1+ ions.

The equation representing the first ionisation energy of sodium is shown below.

$$Na(g) \longrightarrow Na^+(g) + e^- \qquad \text{1st ionisation energy} = +496 \text{ kJ mol}^{-1}$$

Factors affecting ionisation energy

Electrons are held in their shells by attraction from the nucleus. The first electron lost will be from the highest occupied energy level. This electron experiences least attraction from the nucleus.

Three factors affecting the size of this attraction are shown below.

These factors are very important and help to explain many chemical ideas throughout the course.

Learn them!

> **KEY POINT**
>
> **Atomic radius**
> The greater the distance between the nucleus and the outer electrons, the less the attractive force. Attraction falls rapidly with increasing distance and so this factor is very important and has a big effect.
>
> **Nuclear charge**
> The greater the number of protons in the nucleus the greater the attractive force.
>
> **Electron shielding or 'screening'**
> The outer shell electrons are repelled by any inner shells between the electrons and the nucleus.
>
> This repelling effect, called electron shielding or screening, reduces the overall attractive force experienced by the outer electrons.

Evidence for shells

EDEXCEL	M1	OCR	M1
NICCEA	M1	SALTERS	M1
NUFFIELD	M1	WJEC	CH1

> **KEY POINT**
> The **second ionisation energy** of an element is the energy required to remove **1 electron** from each atom in **1 mole** of **gaseous 1+ ions** to form 1 mole of gaseous 2+ ions.

This pattern shows that a sodium atom has 3 shells:

the first shell ($n = 1$) contains 2 electrons

the second shell ($n = 2$) contains 8 electrons

the third shell ($n = 3$) contains 1 electron.

The equation representing the second ionisation energy of sodium is shown below.

$$Na^+(g) \longrightarrow Na^{2+}(g) + e^-$$ 2nd ionisation energy $= +4563$ kJ mol^{-1}

Sodium, with 11 electrons, has 11 successive ionisation energies, shown as the graph below. Successive ionisation energies provide evidence for the different energy levels of the principal quantum numbers.

These **2** electrons are in the 1st shell ($n = 1$), closest to the nucleus. These electrons experience very **strong** attractive forces from the nucleus.

This **1** electron is in the 3rd shell ($n = 3$), furthest from the nucleus. This electron experiences comparatively **weak** attractive forces from the nucleus.

These **8** electrons are in the 2nd shell ($n = 2$). These electrons experience reasonably strong attractive forces from the nucleus.

ionisation energy /kJ mol^{-1}

0 1 2 3 4 5 6 7 8 9 10 11
ionisation number

Evidence for sub-shells

AQA	M1	NUFFIELD	M1
EDEXCEL	M1	OCR	M1
NICCEA	M1	WJEC	M1

The variation in first ionisation energies across a period in the Periodic Table provides evidence for the existence of sub-shells. The graph below shows this variation across Period 2 (Li → Ne)

Further details of trends in 1st ionisation energies are found on p.66.

first ionisation energy

filling of 2s sub-shell

filling of 2p sub-shell

2p sub-shell half fills

pairing of 2p electrons

atomic number

General trend

There is a general rise in the first ionisation energy across a period.

The nuclear charge **increases** giving greater attraction.

There is little variation in atomic radius.

Electrons are added to the **same shell** ($n = 2$) that is the **same order of distance** from the nucleus.

Across a period, increased nuclear charge is most important.

All elements in the period experience the same degree of electron shielding from inner shells.

There is only one inner shell ($n = 1$) for all elements in Period 2.

> **KEY POINT**
> Across a period, the increased **nuclear charge** is the most important factor and the greater attractive force leads to an increased first ionisation energy.

Looking closer at first ionisation energies

Closer analysis provides evidence for sub-shells. Across the period, the graph shows three sections:

- a rise from lithium to beryllium,
- a fall to boron followed by a rise to nitrogen,
- a fall to oxygen followed by a rise to neon.

Comparing beryllium and boron

The fall in 1st ionisation energy from beryllium to boron marks the start of filling the 2p sub-shell which is at higher energy than the 2s sub-shell.

Beryllium versus boron: higher energy level.

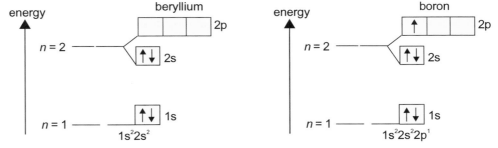

- Beryllium's outermost electron is in the 2s sub-shell.
- Boron's outermost electron is in the 2p sub-shell.
- Boron's outermost electron is easier to remove because the 2p sub-shell has a higher energy level than the 2s sub-shell.

Comparing nitrogen and oxygen

The fall in 1st ionisation energy from oxygen to nitrogen marks the start of electron pairing in the p-orbitals of the 2p sub-shell.

Nitrogen versus oxygen: electron pairing.

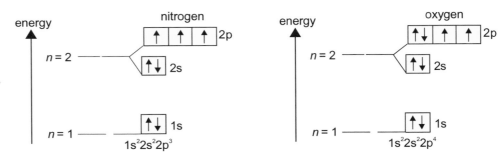

- Both nitrogen and oxygen have their outer electrons in the same 2p sub-shell with the same energy level.
- Oxygen has one 2p orbital with an electron pair – nitrogen, with one electron in each 2p orbital, has unpaired 2p electrons only.
- Because of electron repulsion, it is easier to remove one of oxygen's paired 2p electrons.

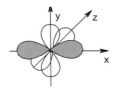

Nitrogen
2p sub-shell half-full
1 electron in each 2p orbital
parallel spins at right angles

Oxygen
2p electrons start to pair
paired electrons repel

27

Sample question and model answer

1

(a) Carbon has three main isotopes, ^{12}C, ^{13}C and ^{14}C. By comparing the atomic structures of these isotopes, explain what is meant by the term *isotopes*.

> Isotopes are atoms of the same element with different masses. ✓
> The three isotopes of carbon have different numbers of neutrons. ✓
> ^{12}C has 6 neutrons, ^{13}C has 7 neutrons and ^{14}C has 8 neutrons. ✓

[3]

When comparing isotopes, it is important to stress the different numbers of neutrons.

(b) The first ionisation energy of carbon is 1086 kJ mol^{-1}.
   (i) Explain the term first ionisation energy.

> The **first** ionisation energy of an element is the energy required to remove 1 **electron** ✓ from each atom in 1 **mole** ✓ of gaseous atoms ✓ to form 1 mole of gaseous 1+ ions.

Here, there will be 1 mark for the equation. A second mark will be available for the state symbols – important for ionisation energies. Don't forget them.

   (ii) Write an equation, including state symbols, to represent the first ionisation energy of carbon.

> $C(g) \rightarrow C^+(g) + e^-$ ✓✓

   (iii) Suggest why there is very little difference between the first ionisation energies of ^{12}C, ^{13}C and ^{14}C.

The word 'suggest' implies that you will need to use your chemical understanding to arrive at a sensible response. There are usually more marks allocated to this type of question than the maximum score possible. Here any 2 of the 3 marking points will secure the 2 marks.

> The isotopes have the same number of protons giving the same nuclear attraction ✓ on the same number of electrons. ✓ The difference in neutrons has no effect on the nuclear attraction. ✓

[7]

(c) Describe and explain the variation in first ionisation energies across Period 2 from lithium to neon.

Note the importance of nuclear charge (for trends across a period).

> Across Period 2, the first ionisation energies increase ✓ because the nuclear charge increases ✓ as electrons are added to the same shell. ✓
> The first ionisation energy of B is less than that of Be. ✓ The outermost electron of B is in the 2p sub-shell, ✓ at a higher energy ✓ than the outermost electron of Be, in the 2s sub-shell.
> The first ionisation energy of O is less than that of N. ✓ The three p electrons in N are unpaired in each of the three orbitals in the p sub-shell. ✓ O has pairing in one p orbital. ✓ The paired electrons in O repel ✓ and are more easily lost.

The emphasis here should be on the LOW ionisation energy of B (2p sub-shell starts) and of O (2p electron pairing starts).

It is NOT Be and N that are out of step. Be and N are NOT 'more stable'.

This is a common cause for the loss of marks in exams.

Here there are 10 marking points for 9 marks maximum.

[9]
[Total: 19 marks]

Practice examination questions

1 Chemists use a model of an atom that consists of sub-atomic particles: protons, neutrons and electrons.

(a) Complete the table below to show the properties of these sub-atomic particles.

particle	relative mass	relative charge
proton		
neutron		
electron		

[3]

(b) The particles in each pair below differ **only** in the number of protons **or** neutrons or electrons. Explain what the difference is within each pair.
   (i) ^{6}Li and ^{7}Li
   (ii) ^{32}S and ^{32}S^{2-}
   (iii) ^{39}K$^+$ and ^{40}Ca^{2+}.

[6]
[Total: 9 marks]

Cambridge Chemistry Foundation Q1 June 1999

2 (a) In terms of the numbers of sub-atomic particles, state **one** difference and **two** similarities between two isotopes of the same element. [3]

(b) Give the chemical symbol, including its mass number, for an atom which has 3 electrons and 4 neutrons. [1]

(c) (i) An element has an atomic number of 16 and it forms an ion with a charge of 2–. State the electronic configuration of this ion.
   (ii) To which block in the Periodic Table does this element belong? [2]

(d) (i) Write an equation for the process involved in the first ionisation energy of aluminium.
   (ii) Explain why the fourth ionisation energy of aluminium is much greater than the third. [6]

[Total: 12]

NEAB Atomic Structure, Bonding and Periodicity Q1 Feb 1996

3 The first ionisation energies of the elements lithium to carbon are given below:

element	Li	Be	B	C
ionisation energy / kJ mol^{-1}	519	900	799	1090

(a) Explain the meaning of the term *first ionisation energy*. [3]

(b) Explain why beryllium has a higher first ionisation energy than lithium. [2]

(c) Write the electronic configuration of a boron atom. [1]

(d) Explain why the first ionisation energy of boron is less than that for beryllium. [2]

(e) Predict a value for the first ionisation energy of nitrogen and explain how you estimated your answer. [2]

[Total: 10]

NEAB Atomic Structure, Bonding and Periodicity Q2 June 1995

Chapter 2

Atoms, moles and equations

The following topics are covered in this chapter:

- *Atomic mass*
- *The mole*

- *Formulae, equations and reacting quantities*

2.1 Atomic mass

After studying this section you should be able to:

- define relative masses, based on the ^{12}C scale
- describe the basic principles of the mass spectrometer
- calculate the relative atomic mass of an element from its isotopic abundance or mass spectrum
- calculate the relative molecular mass of a compound from relative atomic masses

LEARNING SUMMARY

Relative atomic mass

AQA	M1	OCR	M1
EDEXCEL	M1	SALTERS	M1
NICCEA	M1	WJEC	CH1
NUFFIELD	M1		

Carbon-12: an international standard

The mass of an atom is too small to be measured on even the most sensitive balance. Chemists use **relative** masses to compare the atomic masses of different elements. First, we must have an atom with which to compare other atoms and an atom of the **carbon-12 isotope** is chosen as the international standard for the measurement of atomic mass.

Atomic mass unit (amu)

1 atomic mass unit (amu) is the mass of one-twelfth the mass of an atom of the carbon-12 isotope. This provides the base measurement for atomic masses:

1 amu = 1.661×10^{-23} g

- The mass of an atom of carbon-12 is exactly 12 atomic mass units (amu).
- The mass of one-twelfth of an atom of carbon-12 is exactly 1 amu.

All atoms in a pure isotope have the same atomic structure and the same mass. An isotope's *relative* mass is found by comparison with carbon-12.

> **Relative isotopic mass** is the mass of an atom of an isotope compared with one-twelfth the mass of an atom of carbon-12.
>
> KEY POINT

In most chemistry work, it is reasonable to:
- neglect the tiny contribution to atomic mass from electrons;
- take both the mass of a proton and a neutron as 1 amu.

Relative isotopic mass is then simply the mass number of the isotope, e.g. the relative isotopic mass of ^{16}O is 16.

Relative atomic mass

Note that a relative mass has no units. It is simply a ratio of masses and any mass units will cancel.

Most elements consist of a mixture of isotopes, each with a different mass number. To work out the relative atomic mass of an element we must find the **weighted average mass** of the isotopes present from:

- the natural abundances of the isotopes
- the relative isotopic masses of the isotopes.

Examples of relative atomic masses:

H: 1.008;

Cl: 35.45;

Pb: 207.19

> **Relative atomic mass**, A_r, is the weighted average mass of an atom of an element compared with one-twelfth of the mass of an atom of carbon-12.
>
> KEY POINT

Calculating a relative atomic mass

In most chemistry work, the relative isotopic mass can be taken as a whole number.

Naturally occurring chlorine consists of 75% ^{35}Cl and 25% ^{37}Cl.

$\frac{75}{100}$ of chlorine is the ^{35}Cl isotope;

$\frac{25}{100}$ of chlorine is the ^{37}Cl isotope.

The relative atomic mass of chlorine = $\frac{75}{100} \times 35 + \frac{25}{100} \times 37 = 35.5$

Measuring relative atomic masses

A relative atomic mass is measured using a mass spectrometer.

Remember VIADD:

Vaporisation
Ionisation
Acceleration
Deflection
Detection

- A sample of the element is placed into the mass spectrometer and vaporised.
- The sample is bombarded with electrons forming positive ions.
- The positive ions are accelerated using an electric field.
- The positive ions are deflected using a magnetic field.
- Ions of lighter isotopes are deflected more than ions of heavier isotopes. This separates different isotopes.
- The ions are detected to produce a mass spectrum.

You can follow the stages in the diagram below:

Notice that ions are detected in the mass spectrometer. The ions are formed following electron bombardment.

Relative atomic mass from a mass spectrum

A mass spectrum provides:

- the relative isotopic masses of the isotopes in an element
- isotopic abundances.

The diagram below shows how a relative atomic mass can be calculated from a mass spectrum.

Mass spectrum of a copper sample

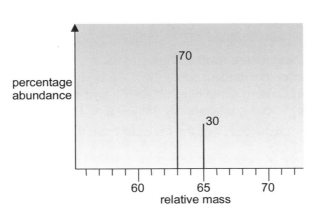

ion	relative mass	percentage abundance
$^{63}Cu^+$	63	70%
$^{65}Cu^+$	65	30%

The relative atomic mass of copper

$= \frac{70}{100} \times 63 + \frac{30}{100} \times 65$

$= 63.6$

Relative masses for compounds

AQA	M1	OCR	M1
EDEXCEL	M1	SALTERS	M1
NICCEA	M1	WJEC	CH1
NUFFIELD	M1		

Relative masses can be used to compare the masses of any chemical species. Much of matter is made up of compounds.

Relative molecular mass

You will find out more about chemical bonding and structure in Chapter 3, p.44.

A simple molecule of Cl_2, H_2O or CO_2 comprises small groups of atoms, held together by chemical bonds.

> **Relative molecular mass, M_r, is the weighted average mass of a molecule of a compound compared with one-twelfth the mass of an atom of carbon-12.**
>
> KEY POINT

The *relative molecular mass* of a compound is found by adding together the relative atomic masses of each atom in a molecule.

Examples of relative molecular masses
Cl_2: $M_r = 35.5 \times 2 = 71$;
H_2O: $M_r = 1.0 \times 2 + 16.0 = 18.0$

For giant structures, the **formula unit** of the compound is the simplest ratio of atoms or ions in the structure.

Some compounds, such as sand SiO_2, exist as 'giant' molecules comprising hundreds of thousands of atoms bonded together. Each molecule in these structures is the size of each crystal and, in a sense, the whole crystal is a molecule.

- Each 'molecule' of SiO_2 has twice as many oxygen atoms as silicon atoms.

Many compounds, such as common salt $NaCl$, are made up of ions and not molecules. These ionic compounds do not have single molecules and form giant structures of oppositely charged ions.

- $NaCl$ has the same number of sodium ions as chloride ions.

Relative formula mass can be used to describe the relative mass of any compound.

Although relative molecular mass is often used with these compounds, a better term to use is relative formula mass.

> **Relative formula mass is the average mass of the formula unit of a compound compared with one-twelfth the mass of an atom of carbon-12.**
>
> KEY POINT

The *relative formula mass* of a compound is found by adding together the relative atomic masses of each atom in the formula.

Examples of relative formula masses
$CaBr_2$: relative formula mass $= 40.1 + 79.9 \times 2 = 199.9$;
Na_3PO_4: relative formula mass $= 23.0 \times 3 + 31.0 + 16.0 \times 4 = 164.0$

Progress check

1 Calculate the relative atomic mass, A_r, of the following elements from their isotopic abundances:
(a) Gallium containing 60% ^{69}Ga and 40% ^{71}Ga.
(b) Silicon containing 92.18% ^{28}Si, 4.71% ^{29}Si and 3.11% ^{30}Si.

2 Use A_r values from the Periodic Table to calculate the relative formula mass of:
(a) CO_2
(b) NH_3
(c) Fe_2O_3
(d) $C_6H_{12}O_6$
(e) $Pb(NO_3)_2$.

1 (a) Gallium: 69.8
(b) Silicon: 28.11.
2 (a) 44
(b) 17
(c) 159.6
(d) 180
(e) 331.

2.2 The mole

After studying this section you should be able to:

- *understand the concept of the mole as the amount of substance*
- *calculate molar quantities using masses, solutions and gas volumes*

LEARNING SUMMARY

The mole

AQA	M1	OCR	M1
EDEXCEL	M1	SALTERS	M1
NICCEA	M1	WJEC	CH1
NUFFIELD	M1		

Counting and weighing atoms

To make a compound such as H_2O, it would be useful to be able to count out atoms. For H_2O, we would need twice as many hydrogen atoms as oxygen atoms. Atoms are far too small to be counted individually but they can be counted using the idea of relative mass:

element	H	C	O
relative mass	1	12	16

For 1 atom of H, C and O, the relative masses are in the ratio 1 : 12 : 16.
For 10 atoms of each, the ratio is 10 : 120 : 160 which is still 1 : 12 : 16.

> Provided we have the same number of atoms of each, the masses are in the ratio of the relative masses.

By measuring masses, chemists are able to actually count atoms.
For 1 g of H, 12 g of C and 16 g of O, the masses are still in the ratio 1 : 12 : 16.
1 g of H contains the same number of atoms as 12 g of C and 16 g of O.

Amount of substance

Chemists use a quantity called *amount of substance* for counting atoms.
Amount of substance is:
- given the symbol n
- measured using a unit called the **mole** (abbreviated as **mol**).

As with relative masses, amount of substance uses the carbon-12 isotope as the standard:
The amount n of carbon-12 atoms in 12 g of ^{12}C is 1 *mol*.

The actual **number** of atoms in 1 mol of carbon-12 is 6.02×10^{23}.

This can be expressed by saying:

'There are 6.02×10^{23} atoms per mole of carbon-12 atoms (or 6.02×10^{23} mol^{-1})'.

6.02×10^{23} mol^{-1} is a constant known as the **Avogadro constant**, L.

> Note that carbon-12 is the **substance** here. So **amount of *substance*** translates to **amount of carbon-12**.

This is a very powerful and useful idea. To make a water molecule from hydrogen and oxygen, we cannot **count** out the atoms but we can **weigh** them out in the correct ratio. To make H_2O, we need 2 atoms of hydrogen for every 1 atom of oxygen which we can get by weighing out 2 g of hydrogen and 16 g of oxygen.

2 g of H contains twice as many atoms as 16 g of O.
2 g of H atoms contains 2 mol of H atoms.
16 g of O atoms contains 1 mol of O atoms.

KEY POINT

Particle

A single particle of a substance may refer to an atom, a molecule, an ion, an electron or to any identifiable particle. Chemists refer to a collection of particles as a **chemical species**.

Amount of substance is not restricted just to atoms. We can have an amount of any chemical species: 'amount of atoms', 'amount of molecules', 'amount of ions', etc.

> **A mole** is the amount of substance that contains as many single particles as there are atoms in exactly 12 g of the carbon-12 isotope.
>
> **KEY POINT**

Using the definition of the mole above, this means that:

- 1 mole of hydrogen atoms, H, contains 6.02×10^{23} hydrogen atoms, H

- 1 mole of oxygen molecules, O_2, contains 6.02×10^{23} oxygen molecules, O_2.

- 1 mole of electrons, e⁻, contains 6.02×10^{23} electrons, e⁻.

It is important to always refer to what the particle is.

For example:

'1 mole of oxygen' could refer to oxygen atoms, O, or to oxygen molecules, O_2.

It is always safest to quote the formula to which the amount of substance refers.

Moles from masses

AQA	M1	OCR	M1
EDEXCEL	M1	SALTERS	M1
NICCEA	M1	WJEC	CH1
NUFFIELD	M1		

Molar mass, *M*

> Molar mass, *M*, is the mass of one mole of a substance.
> The units of molar mass are g mol⁻¹.
>
> **KEY POINT**

For the atoms of an element, the molar mass is simply the relative atomic mass in g mol⁻¹.

- Molar mass of Mg = 24.3 g mol⁻¹
- This simply means that 1 mole of Mg atoms has a mass of 24.3 g.

Molar mass is an extremely useful term that can be applied to any chemical species.

For a compound, the molar mass is found by adding together the relative atomic masses of each atom in the formula:

- Molar mass of CO_2 = 12 + (16 x 2) = 44 g mol⁻¹
- 1 mole of CO_2 has a mass of 44 g.

Examples of molar masses

H   1 g mol⁻¹
H_2O 18 g mol⁻¹
CO_2 44 g mol⁻¹
HNO_3 63 g mol⁻¹

> Amount of substance *n*, mass and molar mass are linked by the expression below:
> $$n = \frac{mass\ (in\ g)}{molar\ mass}$$
>
> **KEY POINT**

In 24 g of C, *amount of C* $= \dfrac{mass\ (in\ g)}{molar\ mass} = \dfrac{24}{12} = 2$ mol

In 11 g of CO_2, *amount of* CO_2 $= \dfrac{mass\ (in\ g)}{molar\ mass} = \dfrac{11}{44} = 0.25$ mol

Progress check

1 (a) What is the amount of each substance (in mol) in the following:
   (i) 124 g P;   (ii) 64 g O;   (iii) 64 g O_2;   (iv) 10 g $CaCO_3$;
   (v) 3.04 g Cr_2O_3?

   (b) What is the mass of each substance (in g) in the following:
   (i) 0.1 mol Na;   (ii) 2.5 mol NO_2;   (iii) 0.3 mol Na_2SO_4;
   (iv) 0.05 mol H_2SO_4;   (v) 0.025 mol Ag_2CO_3?

1 (a) (i) 4 mol;   (ii) 4 mol;   (iii) 2 mol;   (iv) 0.1 mol;   (v) 0.02 mol.
(b) (i) 2.3 g;   (ii) 115 g;   (iii) 42.6 g;   (iv) 4.9 g;   (v) 6.9 g.

Moles from solutions

AQA	M1	OCR	M1
EDEXCEL	M1	SALTERS	M1
NICCEA	M1	WJEC	CH1
NUFFIELD	M1		

> **KEY POINT**
> The concentration of a solution is the amount of solute, in mol, dissolved in 1 dm³ (1000 cm³) of solution.

If the volume of the solution is in dm³:

$$n = c \times V \text{ (in dm}^3\text{)}$$

$n =$ amount of substance, in mol.

$c =$ concentration of solution, in mol dm⁻³.

$V =$ volume of solution, in dm³.

> Dividing a volume measured in cm³ by 1000 converts the volume automatically to dm³.

It is more convenient to measure smaller volumes of solutions in cm³ and the expression becomes:

$$n = c \times \frac{V \text{ (in cm}^3\text{)}}{1000}$$

Molarity

> Molarity refers to the concentration in mol dm⁻³.
>
> Thus 2 mol dm³ and 2 Molar (or 2M) mean the same:
> 2 moles of solute in 1 dm³ of solution.

Example

What is the amount of NaCl (in mol) in 25.0 cm³ of an aqueous solution of concentration 2.00 mol dm⁻³?

$$n(\text{NaCl}) = c \times \frac{V}{1000} = 2 \times \frac{25.0}{1000} = 0.0500 \text{ mol}$$

Standard solutions

Chemists often need to prepare standard solutions with an exact concentration. Using an understanding of the mole, the mass required to prepare such a solution can easily be worked out.

> Concentrations are sometimes referred to in g dm⁻³.
>
> The Na_2CO_3 here has a **molar** concentration of 0.100 mol dm⁻³ and a **mass** concentration of 10.6 g dm⁻³ of Na_2CO_3.

Example

Find the mass of sodium carbonate required to prepare 250 cm³ of a 0.100 mol dm⁻³ solution.

Find the amount of Na_2CO_3 (in mol) required in solution:

$$\text{amount, } n, \text{ of } Na_2CO_3 = c \times \frac{V}{1000} = 0.100 \times \frac{250}{1000} = 0.0250 \text{ mol}$$

Convert moles to grammes

> molar mass of Na_2CO_3 = $23.0 \times 2 + 12.0 + 16.0 \times 3$ = 106.0 g mol⁻¹

$$n = \frac{\text{mass}}{\text{molar mass}} \quad \therefore \quad \text{mass} = n \times \text{molar mass}$$

mass of Na_2CO_3 required = 0.0250×106.0 = 2.65 g

Progress check

1 (a) What is the amount of each substance, in mol, in:
   (i) 250 cm³ of a 1.00 mol dm⁻³ solution;
   (ii) 10 cm³ of a 2.0 mol dm⁻³ solution?

(b) Find the concentration, in mol dm⁻³, for:
   (i) 4 mol in 2 dm³ of solution;
   (ii) 0.0100 mol in 100 cm³ of solution.

(c) Find the concentration, in g dm⁻³, for:
   (i) 2 mol of NaOH in 4 dm³ of solution;
   (ii) 0.500 mol of HNO_3 in 200 cm³ of solution.

1 (a) (i) 0.25 mol; (ii) 0.02 mol.
(b) (i) 2 mol dm⁻³; (ii) 0.1 mol dm⁻³.
(c) (i) 20 g dm⁻³; (ii) 157.5 g dm⁻³.

Moles from gas volumes

AQA	M1	OCR	M1
EDEXCEL	M1	SALTERS	M1
NICCEA	M1	WJEC	CH1
NUFFIELD	M1		

This is sometimes summarised by Avogadro's hypothesis:

'Equal volumes of gases contain the same number of molecules under the same conditions of temperature and pressure.'

For a gas, the amount of gas molecules (in mol) is most conveniently obtained by measuring the gas volume.

Provided that the pressure and temperature are the same, equal volumes of gases contain the same number of molecules.

This means that it does not matter which gas is being measured. By measuring the volume, we are indirectly also counting the number of molecules.

> **KEY POINT**
>
> At room temperature and pressure (r.t.p), 298K (25°C) and 100 kPa, 1 mol of a gas occupies approximately 24 dm^3 = 24 000 cm^3.

At r.t.p., if the volume of the gas is in **dm^3**: $n = \dfrac{V \text{ (in dm}^3)}{24}$

if the volume of the gas is in **cm^3**: $n = \dfrac{V \text{ (in cm}^3)}{24000}$

Example

How many moles of gas molecules are in 480 cm^3 of a gas at r.t.p.?

$$n = \frac{V \text{ (in cm}^3)}{24000} = \frac{480}{24000} = 0.0200 \text{ mol of gas molecules.}$$

The ideal gas equation (AQA and WJEC only)

The conditions will not always be room temperature and pressure. A gas volume depends on temperature and pressure.

The Ideal Gas Equation can be used to find the amount of gas molecules n (in mol) in a volume V (in m^3) at any temperature T (in K) and pressure p (in Pa).

The Ideal Gas Equation: $pV=nRT$ ($R = 8.31$ J K^{-1} mol^{-1})

Before using $pV = nRT$, you must remember to convert any values to Pa, K and m^3.

Conversion rules

cm^3 to m^3	$\times 10^{-6}$
dm^3 to m^3	$\times 10^{-3}$
°C to K	$+ 273$
kPa to Pa	$\times 10^3$

Note that M (molar mass) has units: g mol^{-1}

M_r (relative molecular mass) has no units.

Example

A 0.215 g sample of a volatile liquid, **X**, produces 77.5 cm^3 of gas at 100°C and 100 kPa. Calculate the relative molecular mass of **X**.

Use the ideal gas equation:

$$pV = nRT \qquad \therefore n = \frac{pV}{RT}$$

$$\therefore n = \frac{1.00 \times 10^5 \times 7.75 \times 10^{-5}}{8.31 \times 373} = 0.00250 \text{ mol}$$

Find the molar mass:

$$n = \frac{\text{mass}}{\text{molar mass}}$$

$$\therefore \text{molar mass} = \frac{\text{mass}}{n} = \frac{0.215}{0.00250} = 86.0 \text{ g mol}^{-1}$$

$$\therefore \text{Relative molecular mass } M_r = 86.0$$

Progress check

1 (a) What is the volume, at room temperature and pressure (r.t.p.) of
   (i) 44 g CO_2(g);   (ii) 7 g N_2(g);   (iii) 5.1 g NH_3(g)?
   (b) What is the mass, at r.t.p. of
   (i) 1.2 dm^3 O_2;   (ii) 720 cm^3 CO_2(g);   (iii) 48 cm^3 CH_4(g)?

1 (a) (i) 24 dm^3; (ii) 6 dm^3; (iii) 7.2 dm^3.
   (b) (i) 1.6 g; (ii) 1.32 g; (iii) 0.032 g.

2.3 Formulae, equations and reacting quantities

After studying this section you should be able to:

- understand what is meant by empirical and molecular formula
- calculate the empirical and molecular formula of a substance
- write balanced equations
- use chemical equations to calculate reacting masses and reacting volumes (and vice versa)
- perform calculations involving volumes and concentrations of solutions in simple acid-base titrations

Types of chemical formula

AQA	M1	OCR	M1
EDEXCEL	M1	SALTERS	M1
NICCEA	M1	WJEC	CH1
NUFFIELD	M1		

The **empirical formula** of ethane is CH_3.

Ethane has 1 carbon atom for each of 3 hydrogen atoms.

The molecular formula of ethane is C_2H_6

Each molecule of ethane contains 2 carbon atoms and 6 hydrogen atoms.

Empirical and molecular formula

> **Empirical formula** represents the simplest, whole-number ratio of atoms of each element in a compound.
>
> **Molecular formula** represents the actual number of atoms of each element in a molecule of a compound.

KEY POINT

Formula determination

A formula can be calculated from experimental results using the Mole Concept.

Example 1

Find the empirical formula of a compound formed when 6.75 g of aluminium reacts with 26.63 g of chlorine. [A_r: Al, 27.0; Cl, 35.5.]

Find the molar ratio of atoms:

$$
\begin{array}{ccc}
\text{Al} & : & \text{Cl} \\
= \dfrac{6.75}{27.0} & : & \dfrac{26.63}{35.5} \\
0.25 & : & 0.75 \\
1 & : & 3
\end{array}
$$

Divide by smallest number (0.25): 1 : 3

∴ Empirical formula = $AlCl_3$

Example 2

Find the molecular formula of a compound containing carbon, hydrogen and oxygen only with the composition by mass of carbon: 40.0%; hydrogen: 6.7%. [$M_r = 180$. A_r: H, 1.00; C, 12.0; O, 16.0.]

Percentage of oxygen in compound = $100 - (40.0 + 6.7) = 53.3\%$
100.0 g of the compound contains 40.0 g C, 6.7 g H and 53.3 g O.

Analysis often doesn't give the oxygen content directly but it can easily be calculated.

Find the molar ratio of atoms:

$$
\begin{array}{ccccc}
\text{C} & : & \text{H} & : & \text{O} \\
= \dfrac{40.0}{12.0} & : & \dfrac{6.7}{1.00} & : & \dfrac{53.3}{16.0} \\
= 3.33 & : & 6.7 & : & 3.33 \\
1 & : & 2 & : & 1
\end{array}
$$

Divide by smallest number (3.33): 1 : 2 : 1

∴ Empirical formula = CH_2O

Relate to molecular mass

Each CH_2O unit has a relative mass of $12 + (1 \times 2) + 16 = 30$
M_r of the compound is 180 containing $180/30 = 6$ CH_2O units
∴ molecular formula is $C_6H_{12}O_6$

Progress check

1 Find the empirical formula of the following:
(a) sodium oxide (2.3 g of sodium reacts to form 3.1 g of sodium oxide).
(b) iron oxide (11.16 g of iron reacts to form 15.96 g of iron oxide).

2 Use the following percentage compositions by mass to find the empirical formula of the compound.
(a) A compound of sulphur and oxygen. [S, 40.0%; O, 60.0%]
(b) A compound of iron, sulphur and oxygen. [Fe, 36.8%; S, 21.1%; O, 42.1%]

3 2.8g of a compound of carbon and hydrogen is formed when 2.4 g of carbon combines with hydrogen [M_r: 56]. What is the empirical and molecular formula of the compound formed?

1 (a) Na_2O; (b) Fe_2O_3.
2 (a) SO_3; (b) $FeSO_4$.
3 CH_2; C_4H_8.

Equations

AQA	M1	OCR	M1
EDEXCEL	M1	SALTERS	M1
NICCEA	M1	WJEC	CH1
NUFFIELD	M1		

Chemical reactions involve the **rearrangement** of atoms and ions.

Chemical equations provide two types of information about a reaction:

* qualitative – *which* atoms or ions are rearranging;
* quantitative – *how* many atoms or ions are rearranging.

Balancing equations

The formula of each substance shows the chemicals involved in the reaction.

State symbols can be added to show the physical states of each species under the conditions of the reaction.

State symbols give the physical states of each species in a reaction:

gaseous state, (g);
liquid state, (l);
solid state, (s);
aqueous solution, (aq).

The **qualitative** equation for the reaction of hydrogen with oxygen is shown below:

$$\text{hydrogen} + \text{oxygen} \longrightarrow \text{water}$$
$$H_2(g) + O_2(g) \longrightarrow H_2O(l)$$

In this equation, hydrogen is balanced, oxygen is **not** balanced.

$$H_2(g) + O_2(g) \longrightarrow H_2O(l)$$

hydrogen	2	2	✓
oxygen		2	1	✗

The equation must be **balanced** to give the **same number** of particles of each element on each side of the equation.

Balancing an equation often takes several stages.

* Oxygen can be balanced by placing a '2' in front of H_2O.
* However, this unbalances hydrogen:

$$H_2(g) + O_2(g) \longrightarrow 2\ H_2O(l)$$

hydrogen	2	4	✗
oxygen		2	2	✓

* H can be re-balanced by placing a '2' in front of H_2.
* The equation is now balanced.

Notice that no formula has been changed. You balance an equation by adding numbers in front of a formula only.

$$2\ H_2(g) + O_2(g) \longrightarrow 2\ H_2O(l)$$

hydrogen	4	4	✓
oxygen		2	2	✓

KEY POINT

More balancing numbers.

3 Na_2SO_4 means
Na: $3 \times 2 = 6$;
  S: $3 \times 1 = 3$;
  O: $3 \times 4 = 12$.

With brackets, the subscript applies to everything in the preceding bracket:

2 $Ca(NO_3)_2$ means
Ca: $2 \times 1 = 2$;
  N: $2 \times (1 \times 2) = 4$;
  O: $2 \times (3 \times 2) = 12$.

In a formula, a subscript applies **only** to the symbol immediately preceding,

e.g. NO_2 comprises **1** N and **2** O.

When balancing an equation, you must **not** change any formula. You can only add a balancing number in front of a formula.

The balancing number multiplies everything in the formula by the balancing number,

e.g. $2AlCl_3$ comprises Al: $2 \times 1 = 2$; Cl: $2 \times 3 = 6$.

Progress check

Balance the following equations:
(a) $Li(s) + O_2(g) \longrightarrow Li_2O(s)$
(b) $CH_4(g) + O_2(g) \longrightarrow CO_2(g) + H_2O(l)$
(c) $Al(s) + HCl(aq) \longrightarrow AlCl_3(aq) + H_2(g)$

(a) $4Li(s) + O_2(g) \longrightarrow 2Li_2O(s)$
(b) $CH_4(g) + 2O_2(g) \longrightarrow CO_2(g) + 2H_2O(l)$
(c) $2Al(s) + 6HCl(aq) \longrightarrow 2AlCl_3(aq) + 3H_2(g)$.

Reacting quantities

AQA — M1
EDEXCEL — M1
NICCEA — M1
NUFFIELD — M1
OCR — M1
SALTERS — M1
WJEC — CH1

The balancing numbers give the ratio of the **amount** of each substance, in mol. Using the example above:

equation	$2 H_2(g)$	+	$O_2(g)$	$\longrightarrow$	$2 H_2O(l)$
moles	2 mol		1 mol	$\longrightarrow$	2 mol

An understanding of molar reacting quantities provides chemists with the required recipes to make quantities of chemicals to order.

Chemists use this **quantitative** information to find:
- the *reacting quantities* required to prepare a particular quantity of a product
- the *quantities of products* formed by reacting particular quantities of reactants.

By working out the reacting quantities from a balanced chemical equation, quantities can be adjusted to take into account the required scale of preparation.

Example 1

These examples assume a 100% yield of products. See also percentage yield, page 117.

Calculate the masses of nitrogen and oxygen required to form 3 g of nitrogen oxide, NO (A_r: N, 14; O, 16).

equation	$N_2(g)$	+	$O_2(g)$	$\longrightarrow$	$2 NO(g)$
moles	1 mol	+	1 mol	$\longrightarrow$	2 mol
reacting masses	(14×2) g	+	(16×2) g	$\longrightarrow$	$2 (14 + 16)$ g
	28 g	+	32 g	$\longrightarrow$	60 g
for 3 g NO(g), $\div 20$:	1.4 g	+	1.6 g	$\longrightarrow$	3 g

$\therefore$ 1.4 g of $N_2(g)$ react with 1.6 g of $O_2(g)$ to form 3 g NO(g).

Example 2

You must keep the proportions the same as the reacting quantities.

Any scaling must be applied to all reacting quantities.

Calculate the volume of oxygen, at r.t.p., formed by the decomposition of an aqueous solution containing 5 g of hydrogen peroxide, H_2O_2. [A_r: H, 1; O, 16.]

equation	$2 H_2O_2(aq)$	$\longrightarrow$	$2 H_2O(l)$	+	$O_2(g)$
moles	2 mol	$\longrightarrow$	2 mol	+	1 mol
reacting quantities	$2 [(1 \times 2) + (16 \times 2)]$ g	$\longrightarrow$			1×24 dm^3
	68 g	$\longrightarrow$			24 dm^3
for 1 g H_2O_2, $\div 68$,	1 g	$\longrightarrow$			$\dfrac{24}{68}$ dm^3
for 5 g H_2O_2, $\times 5$,	5 g	$\longrightarrow$			$\dfrac{24}{68} \times 5$ dm^3

$\therefore$ 5 g of H_2O_2 decomposes to form 1.8 dm^3 of $O_2(g)$

Progress check

1 For the reaction: $S(s) + O_2(g) \longrightarrow SO_2(g)$
   (a) How many moles of S and O_2 react to form 0.5 mole of SO_2?
   (b) What mass of SO_2 is formed by reacting 64.2 g of S with O_2?
   (c) What volume of SO_2 forms when 0.963 g of S reacts with O_2 at r.t.p.?

2 Balance the equation: $Mg(s) + O_2(g) \longrightarrow MgO(s)$
   (a) What mass of MgO forms by burning 8.1 g of Mg?
   (b) What volume of O_2 at r.t.p. will react with this mass of Mg?
   (c) What masses of Mg and O_2 are needed to prepare 1 g of MgO?

2 (a) 13.4 g; (b) 4 dm³; (c) 0.6 g Mg, 0.4 g O₂.
1 (a) 0.5 mol S, 0.5 mol O₂; (b) 96.2 g; (c) 0.72 dm³.

Calculations in acid-base titrations

AQA	M1	OCR	M1
EDEXCEL	M1	SALTERS	M2
NICCEA	M1	WJEC	CH2
NUFFIELD	M1		

Calculations for acid-base titrations are best shown with an example:

In a titration, 25.0 cm³ of 0.100 mol dm⁻³ sodium hydroxide NaOH(aq) were found to react exactly with 10.4 cm³ of sulphuric acid, H_2SO_4(aq). Find the concentration of the sulphuric acid.

To solve the problem, we must use the following pieces of information:

> The acid is added from the burette to the alkali.

> An indicator is required to show the 'end-point' when all alkali has reacted with the added acid.

- the balanced equation
$$2NaOH(aq) + H_2SO_4(aq) \longrightarrow Na_2SO_4(aq) + 2H_2O \text{ (l)}$$

- the concentration c_1 and reacting volume V_1 of NaOH(aq)
- the concentration c_2 and reacting volume V_2 of H_2SO_4(aq).

> The concentration and volume of NaOH are known.

From the titration results, the amount of NaOH (in mol) can be calculated:
$$\text{amount of NaOH} = c \times \frac{V}{1000} = 0.100 \times \frac{25.0}{1000} = 0.00250 \text{ mol}$$

> Using the equation determine the number of moles of the second reagent

From the equation, the amount of H_2SO_4 (in mol) can be determined:
$$2 \text{ NaOH (aq)} + H_2SO_4(aq) \longrightarrow Na_2SO_4(aq) + 2H_2O \text{ (l)}$$
2 mol   1 mol *(balancing numbers)*
$\therefore$ 0.00250 mol NaOH reacts with 0.00125 mol H_2SO_4
amount of H_2SO_4 that reacted = 0.00125 mol

The concentration (in mol dm⁻³) of H_2SO_4 can be calculated by scaling to 1000 cm³:

10.4 cm³ H_2SO_4(aq) contains 0.00125 mol H_2SO_4

> Work out the concentration of H_2SO_4(aq), in mol dm⁻³.

1 cm³ H_2SO_4(aq) contains $\frac{0.00125}{10.4}$ mol H_2SO_4

1 dm³ (1000 cm³) H_2SO_4(aq) contains $\frac{0.00125}{10.4} \times 1000 = 0.120$ mol H_2SO_4

$\therefore$ concentration of H_2SO_4(aq) is 0.120 mol dm⁻³

Progress check

1 25.0 cm³ of 0.500 mol dm⁻³ NaOH(aq) react with 23.2 cm³ of HNO_3(aq).
$$HNO_3(aq) + NaOH(aq) \longrightarrow NaNO_3(aq) + H_2O(l)$$
Find the concentration of the nitric acid, HNO_3.

2 25.0 cm³ of 1.00 mol dm⁻³ KOH(aq) react with 26.6 cm³ of H_2SO_4(aq).
$$H_2SO_4(aq) + 2KOH(aq) \longrightarrow K_2SO_4(aq) + 2H_2O(l)$$
Find the concentration of the sulphuric acid, H_2SO_4.

2 0.470 mol dm⁻³.
1 0.539 mol dm⁻³.

Sample question and model answer

The determination of reacting quantities using the Mole Concept is one of the most important concepts in chemistry. This can be tested in exam questions almost anywhere and it underpins much of the content of a chemistry course at this level. To succeed at AS Chemistry, it is essential that you grasp this concept.

A tank contained 4000 dm³ of waste hydrochloric acid. It was decided to neutralise the acid by adding slaked lime, $Ca(OH)_2$.

(a) The concentration of the acid was first determined by titration of a 25.0 cm³ sample against 0.121 M sodium hydroxide of which 32.4 cm³ were required.

(i) Calculate the concentration, in mol dm⁻³, of the hydrochloric acid in the sample.

no. of moles of NaOH $= \dfrac{0.121 \times 32.4}{1000} = 3.92 \times 10^{-3}$ ✓

$NaOH + HCl \longrightarrow NaCl + H_2O$
1 mol NaOH reacts with 1 mol HCl

∴ 3.92×10^{-3} mol NaOH reacts with 3.92×10^{-3} mol HCl ✓
25.0 cm³ HCl(aq) contains 3.92×10^{-3} mol HCl

1 cm³ HCl(aq) contains $\dfrac{3.92 \times 10^{-3}}{25}$ mol HCl

1 dm³ (1000 cm³) HCl(aq) contains $\dfrac{3.92 \times 10^{-3}}{25} \times 1000 = 0.157$ mol HCl.

∴ concentration of HCl(aq) is 0.157 mol dm⁻³ ✓

Watch out for 'scaling'. This is often missed, even by the best candidates. Read all parts of a question thoroughly to see if you have missed anything.

(ii) Calculate the total number of moles of HCl in the tank.
Total number of moles of HCl in the tank $= 4000 \times 0.157 = 628$ ✓

[4]

(b) Calculate the mass, in kg, of slaked lime required to neutralise the acid. Slaked lime reacts with hydrochloric acid according to the equation shown below:

$$Ca(OH)_2 + 2HCl \longrightarrow CaCl_2 + 2H_2O$$

1 mol $Ca(OH)_2$ reacts with 2 mol HCl

∴ $\dfrac{628}{2}$ mol $Ca(OH)_2$ reacts with 628 mol HCl ✓

Notice that the same ideas are used here.

The correct reacting quantities from the equation are essential.

Molar mass of $Ca(OH)_2 = 40 + (16 + 1)2 = 74$ g mol⁻¹ ✓

Mass of $Ca(OH)_2 = \dfrac{628}{2} \times 74 = 23200$ g $= 23.2$ kg (to 3 significant figures) ✓

[3]

(c) The slaked lime was manufactured by roasting limestone and then adding water:

$$CaCO_3 \longrightarrow CaO + CO_2$$
$$CaO + H_2O \longrightarrow Ca(OH)_2$$

Calculate the mass of limestone which is required to produce 1 kg of slaked lime.

It is important to always show full working. If you do make a mistake early on in a calculation, you will only be penalised once if the method that follows is sound.

In 1 kg $Ca(OH)_2$, number of moles of $Ca(OH)_2 = \dfrac{1 \times 10^3}{74} = 13.5$ ✓

From the equations, 1 mol $CaCO_3 \longrightarrow 1$ mol CaO $\longrightarrow 1$ mol $Ca(OH)_2$
∴ number of moles of $CaCO_3 = 13.5$
∴ Mass of $CaCO_3 = 13.5 \times 100 = 13500$ g $= 1.35$ kg ✓

[2]
[Total: 9]

NEAB Q3 Atomic structure, bonding and periodicity Feb 1996 (modified)

Practice examination questions

1 (a) When 0.25 g of sodium metal was added to 200 cm³ (an excess) of water, the following reaction occurred.

$$Na(s) + H_2O(l) \longrightarrow NaOH(aq) + ½H_2(g)$$

   (i) Calculate the number of moles of sodium taking part in the reaction.

   (ii) Calculate the molarity of the sodium hydroxide solution which was formed.

   (iii) Calculate the volume of hydrogen gas produced at r.t.p. (298 K and 100 kPa). Assume that hydrogen is insoluble in water under these conditions.

[6]

   (b) In another experiment 25.0 cm³ of 0.183 M sodium hydroxide were neutralised by 13.7 cm³ of sulphuric acid according to the following equation.

$$2NaOH(aq) + H_2SO_4(aq) \longrightarrow Na_2SO_4(aq) + 2H_2O(l)$$

   Calculate the molarity of the sulphuric acid.

[3]
[Total: 9]

NEAB Atomic Structure, Bonding and Periodicity Q3 Feb 1997 (modified)

2 50 kg of pure sulphuric acid were accidentally released into a lake when a storage vessel leaked. Two methods were proposed to neutralise it.

   (a) The first proposal was to add a solution of 5 M NaOH to the lake. Sodium hydroxide reacts with sulphuric acid as follows:

$$2NaOH(aq) + H_2SO_4(aq) \longrightarrow Na_2SO_4(aq) + 2H_2O(l)$$

   Calculate the volume of 5 M NaOH required to neutralise the sulphuric acid by answering the following questions.

   (i) How many moles of sulphuric acid are there in 50 kg of the acid?

   (ii) How many moles of sodium hydroxide are required to neutralise this acid?

   (iii) Calculate the volume, in dm³, of 5 M NaOH which contains this number of moles.

[4]

   (b) The second proposal was to add powdered calcium carbonate which reacts as follows:

$$CaCO_3(s) + H_2SO_4(aq) \longrightarrow CaSO_4(s) + H_2O(l) + CO_2(g)$$

   Calculate the mass of calcium carbonate required to neutralise 50 kg of sulphuric acid.

[3]

   (c) Suggest **two** reasons why the addition of calcium carbonate to neutralise the acid is the preferred method in practice.

[2]
[Total: 9]

NEAB Atomic Structure, Bonding and Periodicity Q3 Feb 1995

3 Azides are compounds of nitrogen, used mainly as detonators in explosives. However, sodium azide, NaN_3, decomposes non-explosively on heating to release nitrogen gas. This provides a convenient method of obtaining pure nitrogen in the laboratory.

$$2NaN_3(s) \longrightarrow 2Na(l) + 3N_2(g)$$

(a) A student prepared 1.80 dm³ of pure nitrogen in the laboratory by this method. This gas volume was measured at room temperature and pressure (r.t.p.).

   (i) How many moles of nitrogen, N_2, did the student prepare?
   [Assume that 1 mole of gas molecules occupies 24.0 dm³ at r.t.p.]

   (ii) What mass of sodium azide did the student heat? [3]

(b) After cooling, the student obtained 1.15 g of solid sodium. She then carefully reacted this sodium with water to form 25.0 cm³ of aqueous sodium hydroxide:

$$2Na(s) + 2H_2O(l) \longrightarrow 2NaOH(aq) + H_2(g)$$

Calculate the concentration, in mol dm⁻³, of the aqueous sodium hydroxide. [2]

(c) Liquid sodium has been used as a coolant in nuclear reactors.

   (i) Suggest **one** property of sodium that is important for this use.

   (ii) Suggest **one** particular hazard of sodium in this use. [2]

[Total: 7]

Cambridge Chemistry Foundation Q5 March 1998

Chapter 3

Bonding and structure

The following topics are covered in this chapter:

- *Chemical bonding*
- *Shapes of molecules*
- *Electronegativity, polarity and polarisation*

- *Intermolecular forces*
- *Bonding, structure and properties*

3.1 Chemical bonding

After studying this section you should be able to:

- *appreciate that compounds often contain atoms or ions with electron structures of a noble gas*
- *understand the nature of an ionic bond in terms of electrostatic attraction between ions*
- *determine the formula of an ionic compound from the ionic charges*
- *understand the nature of a covalent bond and a dative covalent (coordinate) bond in terms of the sharing of an electron pair*
- *construct 'dot-and-cross' diagrams to demonstrate ionic and covalent bonding*
- *describe metallic bonding*

LEARNING SUMMARY

Why do atoms bond together?

AQA	M1	OCR	M1
EDEXCEL	M1	SALTERS	M1
NICCEA	M1	WJEC	CH1
NUFFIELD	M1		

The Octet Rule

Under normal conditions, the only elements that can exist as single atoms are the Noble Gases in Group 0 of the Periodic Table. The atoms of all other elements are bonded together. To find out why, we need to look at the electronic configurations of the atoms of the Noble Gases.

> Atoms of the Noble Gases are stable because all their electrons are paired.

The Noble Gases get their name from their unreactivity.

In the atoms of a noble gas:

- all electrons are paired,
- the bonding shells are full.

He atom
$1s^2$

Ne atom
$1s^2 2s^2 2p^6$

Ar atom
$1s^2 2s^2 2p^6 3s^2 3p^6$

This arrangement of electrons is particularly stable and is often referred to as a stable 'octet' (as in neon and argon).

> This tendency to acquire a noble gas electron structure is often refer to as the 'Octet Rule'.

Apart from the Noble Gases, atoms are bonded together. Thus the elements oxygen and nitrogen exists as O_2 and N_2. Compounds exist when atoms of different elements bond together as in CO_2, CO and $NaCl$. In all these examples, electrons have been shared, transferred or rearranged in such a way that the atoms of the elements have acquired a noble gas electron structure. If this creates an outer shell of 8 electrons (as in all these cases), the '*Octet Rule*' is obeyed. All these themes are discussed in the sections that follow.

Types of bonding

Chemical bonds are classified into three main types: ionic, covalent and metallic.

- **Ionic** bonding occurs between the atoms of a **metal** and a **non-metal**. For example NaCl, MgO, Fe_2O_3.
- **Covalent** bonding occurs between the atoms of **non-metals.** For example O_2, H_2, H_2O, diamond and graphite.
- **Metallic** bonding occurs between the atoms of **metals**. For example all metals such as iron, zinc, aluminium, etc.; alloys as in brass (copper and zinc), bronze (copper and tin).

> **KEY POINT**
>
> Always decide the type of bonding before:
> - drawing a 'dot-and-cross' diagram
> - predicting how a compound reacts
> - predicting the properties of a compound.

Many mistakes are made in chemistry by confusing the type of bonding.

Covalent and ionic compounds are bonded differently and they behave differently.

You cannot apply 'ionic bonding ideas' to a compound that is covalent.

Ionic bonding

AQA	M1	OCR	M1
EDEXCEL	M1	SALTERS	M1
NICCEA	M1	WJEC	CH1
NUFFIELD	M1		

Ionic bonds

An ionic bond is formed when electrons are **transferred** from a metal atom to a non-metal atom forming **oppositely-charged ions**.

> **KEY POINT**
>
> An ionic bond is the electrical attraction between oppositely charged ions.

Ionic bonds are present in a compound of a metal and a non-metal.

Example: sodium chloride, NaCl

Sodium chloride, NaCl, is formed by transfer of one electron from a sodium atom ($[Ne]3s^1$) to a chlorine atom ($[Ne]3s^23p^5$) with the formation of ions:

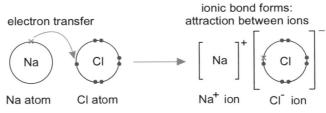

electron transfer

ionic bond forms: attraction between ions

Na atom   Cl atom Na^+ ion Cl^- ion

For showing chemical bonding, 'dot-and-cross' diagrams' often only show the outer shell.

A sodium atom has lost an electron to acquire the electron structure of neon:

$$Na \longrightarrow Na^+ + e^-$$
$$[Ne]3s^1 \longrightarrow [Ne] + e^-$$

A chlorine atom has gained an electron to acquire the electron structure of argon:

$$Cl + e^- \longrightarrow Cl^-$$
$$[Ne]3s^23p^5 + e^- \longrightarrow [Ne]3s^23p^6 \text{ or } [Ar]$$

The ions that are formed have stable noble gas electron structures with full outer electron shells.

Example: magnesium chloride, $MgCl_2$

A magnesium atom ($[Ne]3s^2$) has two electrons in its outer shell. One electron is transferred to each of two chlorine atoms ($[Ne]3s^23p^5$) forming ions.

The Mg^{2+} and Cl^- ions have noble gas electronic configurations.

transfer of 2 electrons

ionic bond forms: attraction between ions

Cl atom   Mg atom   Cl atom Cl^- ion Mg^{2+} ion Cl^- ion

Giant ionic lattices

Each ion is surrounded by oppositely-charged ions, forming a giant ionic lattice.

- Although it is convenient to look at ionic bonding between two ions only, each ion is able to attract oppositely charged ions in all directions.
- This results in a **giant ionic lattice** structure comprising hundreds of thousands of ions (depending upon the size of the crystal).
- This arrangement is characteristic of all ionic compounds.

Part of the sodium chloride lattice

- Na^+
- Cl^-

- Each Na^+ ion surrounds 6 Cl^- ions
- Each Cl^- ion surrounds 6 Na^+ ions

Ionic charges and the Periodic Table

An ionic charge can be predicted from an element's position in the Periodic Table.

The diagram below shows elements in Periods 2 and 3 of the Periodic Table.

group	1	2	3	4	5	6	7	0
number of outer shell electrons	1	2	3	4	5	6	7	8
element	Li	Be	B	C	N	O	F	Ne
ion	Li^+				N^{3-}	O^{2-}	F^-	
element	Na	Mg	Al	Si	P	S	Cl	Ar
ion	Na^+	Mg^{2+}	Al^{3+}		P^{3-}	S^{2-}	Cl^-	

Remember that any metal ions will be positive and non-metal ions negative.

- Metals in Groups 1–3 **lose** sufficient electrons from their atoms to form ions with the electron configuration of the **previous** noble gas;
- Non-metals in Groups 5–7 **gain** sufficient electrons to their atoms to form ions with the electron configuration of the **next** noble gas.
- Be, B, C and Si do not normally form ionic compounds because of the large amount of energy required to transfer electrons forming their ions.

Groups of covalent-bonded atoms can also lose or gain electrons to give ions such as those shown below.

Learn these

1+		1–		2–		3–	
ammonium	NH_4^+	hydroxide	OH^-	carbonate	CO_3^{2-}	phosphate	PO_4^{3-}
		nitrate	NO_3^-	sulphate	SO_4^{2-}		
		nitrite	NO_2^-	sulphite	SO_3^{2-}		
		hydrogen-carbonate	HCO_3^-				

Predicting ionic formulae

Although an ionic compound comprises charged ions, its overall charge is zero.

> In an ionic compound,
> total number of positive charges from positive ions $=$ total number of negative charges from negative ions.

KEY POINT

Working out an ionic formula from ionic charges

calcium chloride:	ion	charge	aluminium sulphate:	ion	charge
equalise charges:	Ca^{2+}	2+	equalise charges:	Al^{3+}	3+
				Al^{3+}	3+
	Cl^-	1–			
	Cl^-	1–		SO_4^{2-}	2–
				SO_4^{2-}	2–
				SO_4^{2-}	2–
total charge must be zero:	0		total charge must be zero:	0	
formula:	$CaCl_2$		formula:	$Al_2(SO_4)_3$	

Covalent bonding

AQA	M1	OCR	M1
EDEXCEL	M1	SALTERS	M1
NICCEA	M1	WJEC	CH1
NUFFIELD	M2		

Covalent bonds

A covalent compound comprises molecules: groups of atoms held together by covalent bonds.

> **KEY POINT**
> A molecule is the smallest part of a covalent compound that can take part in a chemical reaction.

> Covalent bonds are present in a compound of two non-metals.

A covalent bond is formed when electrons are **shared** rather than transferred.

> **KEY POINT**
> A covalent bond is a shared pair of electrons.

Example: hydrogen, H_2

The covalent bond in a hydrogen molecule H_2 forms when two hydrogen atoms ($1s^1$) bond together, each contributing 1 electron to the bond:

In forming the covalent bond, each hydrogen atom in the H_2 molecule has:

> A single covalent bond is written as A–B.

* control of 2 electrons – one of its own and the second from the other atom
* the electron configuration of the noble gas helium.

When a covalent bond forms, each atom *often* acquires a stable noble gas electronic configuration with a full outer bonding shell.

This diagram shows how two atoms are bonded together by a covalent bond.

Unlike an ionic bond, a covalent bond is *directional* and acts solely between the two atoms involved in the bond. Thus hydrogen above exists simply as H_2 molecules.

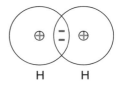

In a hydrogen molecule, the nucleus of each hydrogen atom is attracted towards the electron pair of the covalent bond.

The diagrams below show examples of covalent bonding in some simple molecules.

> Notice that each atom contributes one electron to the covalent bond.

Cl_2

H_2O NH_3

CH_4

Multiple covalent bonds

A covalent bond in which **one** pair of electrons is shared is known as a single bond, e.g. Cl_2, written as Cl–Cl.

Atoms can also share more than one pair of electrons to form a multiple bond.

- Sharing of **two** pairs of electrons forms a double bond, e.g. O_2, written as O=O.
- Sharing of **three** pairs of electrons forms a triple bond, e.g. N_2, written as N≡N.

O_2 N_2

Dative covalent bonds

> A dative covalent bond forms when the shared pair of electrons comes from just **one** of the bonded atoms.

A dative covalent or coordinate bond is one in which **one** of the atoms supplies **both** the shared electrons to the covalent bond. A dative covalent bond is written as A→B, where the direction of the arrow shows the direction in which the electron pair is donated.

Example: ammonium ion, NH_4^+

- The ammonium ion, NH_4^+, forms when ammonia NH_3 bonds with a proton H^+, with both the bonding electrons coming from NH_3.
- Note that the ammonium ion has three covalent bonds and one dative covalent bond:

> The involvement of the electron pair as a dative covalent bond is often shown as an arrow:

> Note that once NH_4^+ has formed, all 4 bonds are equivalent and you cannot tell which was formed by a dative covalent bond.

NH_3 NH_4^+

> In exams, these are easy marks provided that the facts have been learnt.

The Octet Rule doesn't always work

> The Octet Rule is useful in chemistry but it can be broken. It is probably more important that all unpaired electrons are paired.

When covalent bonds form, unpaired electrons often pair up to form a noble gas electronic configuration according to the Octet Rule (see page 44). Many molecules, however, form electronic configurations that do not form an octet. Some molecules have 6, 10, 12 or even 14 electrons in the outer bonding shell. These compounds form by the pairing of unpaired electrons.

Phosphorus forms two chlorides, PCl_3 and PCl_5.

PCl_3 obeys the Octet Rule.

In PCl_5,

- the outer shell of phosphorus has expanded allowing all 5 electrons to be used in bonding
- the octet rule is broken and phosphorus now has 10 electrons in its outer shell:

> In PCl_3, phosphorus is using 3 valence electrons.
>
> In PCl_5, phosphorus is using 5 valence electrons (the same as the Group number of phosphorus).

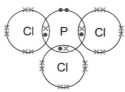

PCl₃: only 3 electrons from the outer bonding shell of phophorus are used in bonding

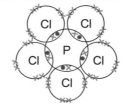

PCl₅: all 5 electrons from the outer bonding shell of phosphorus are used in covalent bonds.

An electron that takes part in forming chemical bonds is called a valence electron.

For elements in the s-block and p-block of the Periodic Table, the maximum number of valence electrons available is usually equal to the group number.

Metallic bonding

AQA	M1	OCR	M1
EDEXCEL	M1	SALTERS	M1
NICCEA	M1	WJEC	CH1
NUFFIELD	M1		

A metallic bond holds atoms together in a solid metal or alloy.

In solid metals, the atoms are ionised.

- The positive ions occupy fixed positions in a lattice.
- The outer shell electrons are delocalised – they are spread throughout the metallic structure and are able to move freely throughout the lattice.

> **KEY POINT**
>
> A metallic bond is the electrostatic attraction between the positive metal ions and delocalised electrons.

> A metallic lattice is held together by electrostatic attraction between positive metal ions and electrons.

In the metallic lattice, each metal atom exists as + ions by releasing its outer valence electrons to the delocalised pool of electrons (often called a 'sea of electrons').

The 'sea of electrons'

Delocalised and localised electrons

The delocalised electrons in metals are spread throughout the metal structure. The delocalisation is such that electrons are able to move throughout the structure and it is impossible to assign any electron to a particular positive ion.

In a covalent bond, the localised pair of electrons is always positioned between the two atoms involved in the bond. The electron charge is much more concentrated.

Progress check

1 Draw *'dot-and-cross'* diagrams of (a) MgO; (b) Na_2O.

2 Using the information on page 46, predict formulae for the following ionic compounds:
   (a) lithium chloride
   (b) potassium sulphide
   (c) lithium nitride
   (d) aluminium oxide
   (e) sodium carbonate
   (f) calcium hydroxide
   (g) aluminium sulphate
   (h) ammonium phosphate.

3 Draw *'dot-and-cross'* diagrams for
   (i) C_2H_6;   (ii) HCN;   (iii) H_3O^+.

2 (a) LiCl; (b) K_2S; (c) Li_3N; (d) Al_2O_3; (e) Na_2CO_3; (f) $Ca(OH)_2$; (g) $Al_2(SO_4)_3$; (h) $(NH_4)_3PO_4$.

3.2 Shapes of molecules

After studying this section you should be able to:

- use electron-pair repulsion to explain the shapes and bond angles of simple molecules
- understand that lone pairs have a larger repulsive effect than bonded pairs leading to distortion of the molecular shape
- understand how multiple bonds affect the shape of a molecule

Electron-pair repulsion theory

AQA	M1	OCR	M1
EDEXCEL	M1	SALTERS	M1
NICCEA	M1	WJEC	CH1
NUFFIELD	M2		

Electron-pair repulsion theory states that:

- the shape of a molecule depends upon the number of electron pairs surrounding the central atom
- electron pairs repel one another and move as far apart as possible.

The shape of any molecule can be predicted by applying this theory.

Molecules with bonded pairs

Learn these shapes and bond-angles.

You can predict the shape of a molecule from its 'dot-and-cross' diagram. Try to relate the number of electron pairs in the examples below with the three-dimensional shape of each molecule.

molecule	$BeCl_2$	BF_3	CH_4	PCl_5	SF_6
electron pairs around central atom	2	3	4	5	6
shape and bond angle	Cl—Be—Cl 180°	B 120°	C 109.5°	Cl—P 90° 120°	S 90°
name of shape	linear	trigonal planar	tetrahedral	trigonal bipyramid	octahedral

Molecules with lone pairs

Each of the examples below are molecules with four electron pairs surrounding the central atom. The molecular shape will therefore be based upon a tetrahedron. However, a lone pair is closer to an atom than a bonded pair of electrons and has a larger repulsive effect than a bonded pair.

> The relative magnitudes of electron-pair repulsions are:
>
> lone-pair/lone-pair > bonded-pair/lone-pair > bonded-pair/bonded-pair.

Lone pairs distort the shape of a molecule and reduce the bond angle. Look at the examples below and notice how each lone pair decreases the bond-angle by 2.5°.

Always draw 3-D shapes – you may be penalised in exams otherwise.

molecule	CH_4	NH_3	H_2O
'dot-and-cross' diagram			
number of lone pairs	0	1	2
shape and bond angle	109.5°	107°	104.5°
name of shape	tetrahedral	pyramidal	non-linear

Molecules with double bonds

In molecules containing multiple bonds, each double bond is treated in the same way as a bonded pair. In the diagram of carbon dioxide below, each double bond is treated as a *'bonding region'*.

molecule	'dot-and-cross' diagram	number of bonding regions	shape and bond angle	name of shape
CO_2		2	180° O=C=O	linear

Progress check

1 For each of the following molecules predict the shape and bond angles.
  (a) BeF_2
  (b) $AlCl_3$
  (c) SiH_4
  (d) H_2S
  (e) PH_3
  (f) CS_2
  (g) SO_3
  (h) SO_2

1 (a) linear, 180°
(b) trigonal planar, 120°
(c) tetrahedral, 109.5°
(d) non-linear, 104.5°
(e) pyramidal, 107°
(f) linear, 180°
(g) trigonal planar, 120°
(h) non-linear, 120°.

3.3 Electronegativity, polarity and polarisation

After studying this section you should be able to:

- *appreciate that many chemical bonds have bonding intermediate between ionic and covalent bonding*
- *describe electronegativity in terms of attraction for bonding electrons*
- *understand the nature of polarity in molecules of covalent compounds*
- *understand the nature of polarisation in ionic compounds*

Ionic or covalent?

AQA	M1	OCR	M1
EDEXCEL	M1	SALTERS	M2
NICCEA	M1	WJEC	CH1
NUFFIELD	M2		

Between the extremes of ionic and covalent bonding, there is a whole range of intermediate bonds, which have both ionic and covalent contributions.

An ionic bond with 100% ionic character would require the complete transfer of an electron from a metal atom to a non-metal atom. In practice, this never completely happens, but a compound such as caesium fluoride Cs^+F^- forms an ionic bond in which the electron transfer is nearly complete with close to 100% ionic character.

In a hydrogen molecule H_2, the two hydrogen atoms are identical and form a covalent bond with 100% covalent character by equally sharing the bonded pair of electrons. However, in molecules such as HCl, the bonded pair of electrons is not shared equally.

An ionic bond is often formed with some covalent character:
- the electron transfer is incomplete
- there is a degree of electron sharing.

A covalent bond is often formed with some ionic character:
- the electrons are not equally shared
- there is a degree of electron transfer.

The sections that follow discuss the factors that lead to intermediate bonding.

Electronegativity

AQA	M1	OCR	M1
EDEXCEL	M1	SALTERS	M2
NICCEA	M1	WJEC	CH1 ·
NUFFIELD	M2		

The nuclei of the atoms in a molecule attract the electron pair in a covalent bond.

> Electronegativity is a measure of the attraction of an atom in a molecule for the pair of electrons in a covalent bond.

The most electronegative atoms attract bonding electrons most strongly.

- In general, small atoms are electronegative atoms.
- The most electronegative atoms are those of highly reactive non-metallic elements (such as O, F and Cl).
- Reactive metals (such as Na and K) have the least electronegative atoms.

How is electronegativity measured?

Notice that the Noble Gases are not included. Although neon and helium atoms are smaller than those of fluorine, they do not form bonds and so they have no affinity for bonded electrons.

See how the Pauling values of electronegativity relate to the element's position in the Periodic Table.

The 'Pauling scale' is often used to compare the electronegativities of different elements. The diagram below shows how the electronegativity of an element relates to its position in the Periodic Table. The numbers give the Pauling electronegativity values.

electronegativity increases						
Li	Be	B	C	N	O	F
1.0	1.5	2.0	2.5	3.0	3.5	4.0
Na						Cl
0.9						3.0
K						Br
0.8						2.8

Fluorine, the most electronegative element has:

- an electronegativity of 4.0
- small atoms with a greater attraction than larger atoms for the pair of electrons in a covalent bond.

> **KEY POINT**
>
> The greater the **difference** between electronegativities, the greater the **ionic** character of the bond.
>
> The greater the **similarity** in electronegativities, the greater the **covalent** character of the bond.

Polar and non-polar molecules

AQA	M1	OCR	M1
EDEXCEL	M1	SALTERS	M2
NICCEA	M1	WJEC	CH1
NUFFIELD	M2		

Non-polar bonds

A covalent bond is non-polar when:

- the bonded electrons are shared **equally** between both atoms
- the bonded atoms are the same
- the bonded atoms have similar electronegativities.

When bonding atoms are the same, the attraction for the bonded pair of electrons is the same and the bond is non-polar.

A covalent bond **must** be non-polar if the bonded atoms are the same, as in molecules of H_2 and Cl_2 shown here:

H_2 molecule Cl_2 molecule

Polar bonds

A covalent bond is polar when:

- the electrons in the bond are shared **unequally** making a *polar bond*
- the bonded atoms are different, each has a different electronegativity.

In a polar covalent bond, the bonded electrons are attracted towards the more electronegative of the atoms.

The bond in hydrogen chloride, H–Cl, is predominantly covalent with one pair of electrons shared between the two atoms. However, the chlorine atom is **more electronegative** than the hydrogen atom. The chlorine atom has a greater attraction for the bonded pair of electrons than the hydrogen atom making the covalent bond polar.

The electronegativities of hydrogen halides are discussed in more detail in p.76.

bonded pair of electrons
attracted closer to chlorine

The molecule is *polarised* with a small positive charge δ+ on the hydrogen atom and a small negative charge δ– on the chlorine atom.

The hydrogen chloride molecule is *polar* with a *permanent dipole*.

Symmetrical and unsymmetrical molecules

If the molecule is symmetrical, any dipoles will cancel and the molecule will not have a permanent dipole. The diagram below shows why the symmetrical molecule, CCl_4, is non-polar.

- Each C–Cl bond is polar but the dipoles act in different directions.
- The overall effect is for the dipoles to cancel each other.
- ∴ CCl_4 is a non-polar molecule.

Dipoles in symmetrical molecules cancel.

CCl_4 is non-polar

Although each C–Cl bond is polar, the dipoles cancel

53

Polarisation in ionic compounds

AQA · M1 OCR · M1
EDEXCEL · M2 WJEC · CH1

A 100% ionic bond would only form by complete transfer of electrons from a metallic atom to a non-metallic atom. Many ionic compounds have a degree of covalency resulting from incomplete transfer of electrons.

This effect, called *polarisation*, occurs in ionic compounds with:

- a small + ion (cation) with a high charge density
- a large – ion (anion) with a low charge density.

Cations and Anions

During electrolysis, the negative electrode or *cathode* attracts positive ions, called **cations**.

The positive electrode or *anode* attracts negative ions, called **anions**.

Charge/size ratio increases

Polarisation is most apparent with a small, densely charged 3+ cation.

Polarisation will be greatest when a small, highly-charged cation polarises a large anion.

The electric field around a **small cation** distorts the electron shells around a **large anion**. As a result, electrons are attracted towards the cation, giving a degree of covalency (electron sharing).

$$\text{charge density} = \frac{\text{charge}}{\text{ionic radius}}$$

Notice how the charge density increases as the ionic charge increases:

ion	Na^+	Mg^{2+}	Al^{3+}
charge	1+	2+	3+
radius/nm	0.102	0.072	0.053
charge density	10	28	57

Note the polarisation between Al^{3+} and O^{2-} – a 'perfect' ionic compound would contain spherical ions. Here, the O^{2-} ion is distorted.

Small Al^{3+} ion is highly charged.
O^{2-} ion is distorted and polarised resulting in partial covalent character.

Progress check

1 Decide whether each molecule is polar and state which atom, if any, has the $\delta-$ charge.
   (a) Br_2; (b) H_2O; (c) O_2; (d) HBr; (e) NH_3.

2 (a) Predict the shape of a molecule of BF_3 and of PF_3.
   (b) Explain why BF_3 is non-polar whereas PF_3 is polar.

1 (a) non-polar; (b) polar, $O^{\delta-}$; (c) non-polar; (d) polar, $Br^{\delta-}$; (e) polar, $N^{\delta-}$.

2 (a) BF_3: trigonal planar; PF_3: pyramidal.
   (b) BF_3 has a symmetrical shape. Although each bond is polar ($F^{\delta-}$), the dipoles cancel. PF_3 is not symmetrical. The F atoms are all on the same side of the molecule resulting in a permanent dipole.

3.4 Intermolecular forces

After studying this section you should be able to:

- *understand the nature of van der Waals' forces (induced dipole-dipole interactions)*
- *understand the nature of permanent dipole-dipole interactions and hydrogen bonds*
- *describe the anomalous properties of water arising from hydrogen bonding.*

Van der Waals' forces

AQA	M1	OCR	M1
EDEXCEL	M1	SALTERS	M2
NICCEA	M1	WJEC	CH1
NUFFIELD	M2		

Van der Waals' forces (induced dipole-dipole interactions) exist between all molecules whether polar or non-polar.

> Induced dipole-dipole interactions are often referred to as van der Waals' forces.

What causes van der Waals' forces?

Van der Waals' forces are:

- weak intermolecular forces
- caused by attractions between very small dipoles in molecules.

The diagram below shows how these forces arise between single atoms in a noble gas.

> Ionic and covalent bonds are of comparable strength.
> Intermolecular forces are far weaker.

Movement of electrons produces an oscillating dipole

dipole oscillates and continually changes with time

> An oscillating or instantaneous dipole is caused by an uneven distribution of electrons at an instant of time.

Oscillating dipole induces a dipole in a neighbouring molecule which is induced onto further molecules

induced dipoles attract one another

What affects the strength of van der Waals' forces?

Van der Waals' forces result from interactions of electrons between molecules.

The greater the number of electrons in each molecule:

- the larger the oscillating and induced dipoles
- the greater the attractive forces between molecules
- the greater the van der Waals' forces.

> Van der Waals' forces increase in strength with increasing number of electrons

This can be seen by comparing the boiling points of the Noble Gases.

Noble gas	b. pt./°C	number of electrons	trend
He	−269	2	• easier to distort electron clouds
Ne	−246	10	
Ar	−186	18	• induced dipoles increase
Kr	−152	36	
Xe	−107	54	• boiling point increases
Rn	− 62	86	

Permanent dipole-dipole interactions

AQA	M1	OCR	M1
EDEXCEL	M1	SALTERS	M2
NICCEA	M1	WJEC	CH1
NUFFIELD	M2		

Permanent dipole-dipole interactions are 'non-directional'.

They are simply weak attractions between dipole charges on different molecules.

The small δ+ and δ- charges on a polar molecule attract oppositely charged dipoles on another polar molecule.

This gives a weak *intermolecular force* called a permanent dipole-dipole interaction.

Example: Intermolecular forces between HCl molecules

Between hydrogen chloride molecules, there will be both:

- van der Waals' forces and
- permanent dipole-dipole interactions.

Although both are weak forces, the permanent dipole-dipole interactions are still stronger than the van der Waals' forces.

Hydrogen bonds

AQA	M1	OCR	M1
EDEXCEL	M1	SALTERS	M2
NICCEA	M1	WJEC	CH1
NUFFIELD	M2		

A hydrogen bond is a special type of permanent dipole-dipole interaction found between molecules containing the following groups:

A hydrogen bond is a comparatively strong intermolecular attraction between:

- an electron deficient hydrogen atom, $H^{\delta+}$, on one molecule and
- a lone pair of electrons on a highly electronegative atom of F, O or N on another molecule.

Hydrogen bonding occurs between molecules such as H_2O:

Note the role of the lone pair – this is essential in hydrogen bonding.

A hydrogen bond is shown between molecules as a dashed line.

hydrogen bond formed by attraction between dipole charges on different molecules

Hydrogen bonding is especially important in organic compounds containing –OH or –NH bonds: e.g. alcohols, carboxylic acids, amines, amino acids.

Special properties of water arising from hydrogen bonding

A hydrogen bond has only one-tenth the strength of a covalent bond. However, hydrogen bonding is strong enough to have significant effects on physical properties, resulting in some unexpected properties for water.

The solid (ice) is less dense than the liquid (water)

Solids are usually denser than liquids – but ice is less dense than water.

- Particles in solids are usually packed closer together than in liquids.
- Hydrogen bonds hold water molecules apart in an open lattice structure.

Therefore ice is less dense than water.

When water changes state, the covalent bonds between the H and O atoms in an H_2O molecule are strong and do **not** break.

It is the intermolecular forces that break.

The diagram below shows how the open lattice of ice collapses on melting.

tetrahedral open lattice in ice

hydrogen bonds break

MELTING OF ICE

ice lattice collapses: molecules move closer together

Ice has a relatively high melting point, and water a relatively high boiling point

- There are relatively strong hydrogen bonds between H_2O molecules.
- The hydrogen bonds are extra forces over and above van der Waals' forces.
- These extra forces result in higher melting and boiling points than would be expected from just van der Waals' forces.
- When the ice lattice breaks, hydrogen bonds are broken.

Other properties

The extra intermolecular bonding from hydrogen bonds also explains the relatively high surface tension and viscosity in water.

Progress check

1 Which of the following molecules have hydrogen bonding: H_2O, H_2S, CH_4, CH_3OH, NO_2?

2 Draw diagrams showing hydrogen bonding between:
   (a) 2 molecules of ammonia
   (b) 1 molecule of water and 1 molecule of ethanol.

1 H_2O, CH_3OH.

2 (a)

(b)

3.5 Bonding, structure and properties

After studying this section you should be able to:

- describe the typical properties of an ionic compound in terms of its structure
- describe the typical properties of a covalent compound in terms of simple molecular and giant molecular structures
- describe the structure and associated properties of diamond and graphite
- understand the properties of metals in terms of metallic bonding

Bonds and forces

AQA	M1	OCR	M1
EDEXCEL	M1	SALTERS	M2
NICCEA	M1	WJEC	CH1
NUFFIELD	M2		

These are very important and are needed to understand the links between bonding, structure and properties.

Bonding, structure and properties are all related.

A structure is held together by bonds and forces. The different types of bonds introduced in the last section are shown below.

> Covalent bonds act between atoms.
> Ionic bonds act between ions.
> Metallic bonds act between positive ions and electrons.
> Hydrogen bonds act between polar molecules.
> Dipole-dipole interactions act between polar molecules.
> Van der Waals' forces act between molecules.

KEY POINT

Ionic, covalent and metallic bonds are of comparable strength.

Intermolecular forces are much weaker, and their relative strengths are compared with the strength of covalent bonds in the table below.

type of bond	bond enthalpy / kJ mol⁻¹
covalent bond	200–500
hydrogen bond	5–40
van der Waals' forces	~2

The properties of a substance depend upon its bonding and structure.

Properties of ionic compounds

AQA	M1	OCR	M1
EDEXCEL	M1	SALTERS	M2
NICCEA	M1	WJEC	CH1
NUFFIELD	M1		

Ionic compounds are solids at room temperature.

The strong forces between ions result in high melting and boiling points.

Ionic compounds conduct only when ions are free to move – when molten or in aqueous solution.

Ionic compounds form giant ionic lattices with each ion surrounded by ions of the opposite charge. The ions are held together by strong electrostatic attraction between positive and negative ions. (See page 46.)

High melting point and boiling point
- High temperatures are needed to break the strong electrostatic forces holding the ions rigidly in the solid lattice.

Therefore ionic compounds have high melting and boiling points.

Electrical conductivity
*In the **solid** lattice,*
- the ions are in a fixed position and there are no mobile charge carriers.

Therefore an ionic compound is a non-conductor of electricity in the solid state.

*When **melted** or **dissolved** in water,*
- the solid lattice breaks down,
- the ions are now free to move as mobile charge carriers.

Therefore an ionic compound is a conductor of electricity in liquid and aqueous states.

Solubility

- The ionic lattice often dissolves in polar solvents (e.g. water).
- Polar water molecules break down the lattice and surround each ion in solution as shown below for sodium chloride.

The structure of an aqueous cation such as $Na^+(aq)$ is often shown as a *complex ion*, e.g. $[Na(H_2O)_6]^+$.

- Water molecules attract Na^+ and Cl^- ions.
- Lattice breaks down as it dissolves.
- Water molecules surround ions.

Na^+ attracts $\delta-$ charges on the O atoms of water molecules.

Cl^- attracts $\delta+$ charges on the H atoms of water molecules.

Properties of covalent compounds

AQA	M1	OCR	M1
EDEXCEL	M1	SALTERS	M2
NICCEA	M1	WJEC	CH1
NUFFIELD	M2		

Elements and compounds with covalent bonds have either of two structures:
- a simple molecular structure
- a giant molecular structure.

Simple molecular structures, e.g. iodine, I_2

Intermolecular forces are weak

Only the intermolecular forces break when a simple molecule melts or boils

A common mistake in exams is to confuse intermolecular forces with covalent bonds.

Simple molecular structures form solid lattices with small molecules, such as Ne, H_2, O_2, N_2, held together by weak intermolecular forces.

Low melting point and boiling point

- Low temperatures provide sufficient energy to break the weak intermolecular forces between molecules.

Therefore simple molecular structures have low melting and boiling points.

The simple molecular structure of solid I_2

Giant structure:

high melting point.

Simple molecular structure:

low melting point.

- When the I_2 lattice breaks down, only the weak van der Waals' forces between the I_2 molecules break.
- In the I_2 molecule, the covalent bond, I–I, is strong and does **not** break when the lattice breaks down.

Strong covalent bonds hold together each I_2 molecule

Weak van der Waals' forces between I_2 molecules

Electrical conductivity

- There are no free charged particles.

Therefore simple molecular structures are non-conductors of electricity.

Solubility

- Van der Waals' forces form between a simple molecular structure and a non-polar solvent, such as hexane. These weaken the structure.

Therefore simple molecular structures are often soluble in non-polar solvents (e.g. hexane).

Giant molecular structures, e.g. carbon (diamond and graphite)

Diamond, graphite and SiO_2 are examples of giant molecular lattices.

Giant molecular structures have thousands of atoms bonded together by strong covalent bonds. This type of structure is known by a variety of names: a giant molecular lattice, a giant covalent lattice or a giant atomic lattice.

High melting point and boiling point

Covalent bonds are strong.

The covalent bonds break when a giant molecular structure melts or boils – this only happens at high temperatures.

- High temperatures are needed to break the strong covalent bonds in the lattice.

Therefore giant molecular structures have high melting and boiling points.

Electrical conductivity

- Except for graphite (see below), there are no free charged particles.

Therefore giant molecular structures are non-conductors of electricity.

Solubility

- The strong covalent bonds in the lattice are too strong to be broken by either polar or non-polar solvents.

Therefore giant molecular structures are insoluble in polar *and* non-polar solvents.

Comparison between the properties of diamond and graphite

property	diamond	graphite
structure	symmetrical structure held together by strong covalent bonds throughout lattice *tetrahedral*	strong layer structure but with weak bonds between the layers *hexagonal layers*
electrical conductivity	*poor* • There are no delocalised electrons as all outer shell electrons are used for covalent bonds.	*good* • Delocalised electrons between layers. • Electrons are free to move parallel to the layers when a voltage is applied.
hardness	*hard* • Tetrahedral shape enables external forces to be spread out throughout the lattice.	*soft* • Bonding within each layer is strong but weak forces between layers easily allow layers to slide.

Properties of metals

AQA M1 OCR M1
EDEXCEL M1 SALTERS M2
NICCEA M1 WJEC CH1
NUFFIELD M1

Metals have a giant metallic lattice structures held together by strong electrostatic attractions between positive ions and negative electrons.

High melting point and boiling point

- Generally high temperatures are needed to separate the ions from their rigid positions within the lattice.

Therefore most metals have high melting and boiling points.

Good thermal and electrical conductivity

When a metal conducts electricity, only the electrons move.

- The existence of mobile, delocalised electrons allows metals to conduct heat and electricity well, even in the solid state.
- The electrons are free to flow between positive ions.
- The positive ions do not move.

drift of delocalised electrons across potential difference

Summary of properties from structure and bonding

structure	bonding	melting pt / boiling pt	reason	electrical conductivity	reason	solubility	reason
giant ionic	ionic bonds	high	strong electrostatic attraction, between oppositely charged ions	poor when solid — good when aqueous or molten	ions in a fixed position in lattice — lattice has broken down: mobile ions	good in polar solvents, e.g. water	attraction between ionic lattice and polar solvent
simple molecular	covalent bonds within the molecules	low	weak van der Waals' forces between molecules	poor	no mobile charged particles (electrons or ions)	good in non-polar solvents	van der Waals' forces between molecular structure and solvent
giant molecular	covalent bonds	high	strong covalent bonds between atoms	poor	no mobile charged particles (electrons or ions)	poor	forces within lattice too strong to be broken by solvents
hydrogen bonded	hydrogen bonds	low *but* higher than expected	dipole-dipole attractions between molecules	poor	no mobile charged particles (electrons or ions)	good in polar solvents, e.g. water	attraction between dipoles
giant metallic	metallic bonds	usually high	strong electrostatic attractions between ions and electrons	good	mobile electrons, even in solid state	poor	forces within lattice too strong to be broken by solvents

> A high melting point is the result of any giant structure, bonded together with strong forces.
>
> Giant structures can be ionic, covalent or metallic.

KEY POINT

Progress check

1 For (a) MgO; (b) CH$_4$; (c) SiO$_2$, state the structure and explain the following physical properties: melting and boiling points, electrical conductivity, solubility.

1 (a) giant ionic: high m.pt/b.pt. – strong forces between ions.
Non-conductor when solid – ions fixed in lattice. Conductor when molten or dissolved in water – ions are able to move.
Soluble in water – dipole on water attracted to ions.
(b) simple molecular: low m. pt/b. pt. – weak van der Waals' forces between molecules.
Non-conductor – no mobile charge carriers, electrons localised in covalent bonds.
Soluble in non-polar solvents – van der Waals' forces in solvent interact with van der Waals' forces in CH$_4$.
(c) giant molecular: high m. pt/b. pt. – strong covalent bonds between atoms.
Non-conductor – no mobile charge carriers, electrons localised in covalent bonds.
Insoluble in all solvents – strong covalent bonds are too strong to be broken by either polar or non-polar solvents.

Sample question and model answer

Chemical bonding helps to explain different properties of materials.

(a) Using suitable diagrams, explain what is meant by *ionic*, *covalent* and *metallic* bonding.

> An ionic bond is the electrostatic attraction between ions. ✓
> An example is sodium chloride:

✓ (dot and cross)
✓ (charges on ions)

> A covalent bond is a **shared** ✓ **pair** of electrons. ✓
> An example is hydrogen chloride.

✓ (dot and cross)

> Metallic bonding occurs between positive centres/ions ✓
> surrounded by mobile or delocalised electrons: ✓

> Metallic bonding is the attraction between positive ions and mobile electrons. ✓

The phrase 'electrostatic attraction between ions' is essential when describing an ionic bond.

DO use diagrams in your answers. Here, dot and cross diagrams are essential for showing ionic and covalent compounds

The word 'pair' is essential when describing a covalent bond. Omit it and you risk losing the mark.

There are plenty of examples. Keep them simple and make sure that you choose correctly. Don't use NaCl as a covalent example – it is ionic!

Many candidates do actually make this mistake.

(b) Three substances have ionic, covalent and metallic bonding respectively. Compare and explain the electrical conductivity of the three materials.

> The ionic compound does not conduct electricity when solid ✓
> because the ions are fixed ✓ in a lattice.
> The ionic compound does conduct electricity when aqueous or molten ✓
> because the ions are mobile. ✓
> The covalent compound does not conduct ✓ at all because there are no
> free charge carriers ✓ (electrons or ions).
> The metallic compound does conduct electricity ✓ because the
> delocalised electrons are able to move ✓ across a potential difference.

Special care needed throughout with language. A-grade students will score all of these marks.

Ionic bonding: positive ions and negative ions – **both** carry electricity.

Metallic bonding: positive ions and delocalised electrons – **only** the electrons move and carry electricity.

17 marking points ⟶ maximum of [15]

Practice examination questions

1 (a) An ammonia molecule comprises nitrogen and hydrogen atoms held together by covalent bonds. What is meant by the term *covalent bond*? [2]

(b) Use the formation of the ammonium ion from ammonia to give the meaning of the term *co-ordinate bond*. [2]

(c) Give the bond angle in the ammonium ion and predict, with an explanation, an appropriate value for the bond angle in the ammonia molecule. [3]

(d) Name the major force of attraction existing between molecules in liquid ammonia and explain how this type of force arises. [3]

[Total: 10]

NEAB Atomic Structure, Bonding and Periodicity Q5 Feb 1995 (modified)

2 Describe how the principle of *electron pair repulsion* helps to explain the shapes of simple molecules. Using diagrams, illustrate how this determines the molecular shapes and bond angles in molecules of (a) CH_4; (b) NH_3; (c) BF_3; (d) SF_2. [17]

[Total: 17]

Cambridge Chemistry Foundation Q1 June 1995 (modified)

3 A crystal of iodine, heated gently in a test tube, gave off a purple vapour. A crystal of sodium chloride, heated to the same temperature, remained unchanged.

(a) Name the type of bonding or interaction which occurs:

(i) between iodine atoms in a molecule of iodine;

(ii) between iodine molecules in a crystal of iodine. [2]

(b) Give the formula for the species that is responsible for the purple colour in the vapour. [1]

(c) Explain why iodine vapourises at a relatively low temperature. [1]

(d) Explain, in terms of its bonding, why the sodium chloride crystal did not melt or vapourise. [2]

(e) Describe what happens to the particles in sodium chloride when the solid is heated above room temperature but below its melting point. [1]

(f) Explain why the heat energy required to vapourise sodium chloride (171 kJ mol^{-1}) is greater than the heat energy required to melt it (29 kJ mol^{-1}). [2]

[Total: 9]

NEAB Atomic Structure, Bonding and Periodicity Q4 Feb 1996

4 Suggest reasons for the different boiling points of (a) water, (b) trichloromethane and (c) argon (given below) by considering the nature and strength of the intermolecular forces in each case.

liquid	H_2O	$CHCl_3$	Ar
boiling point /°C	100	62	−186

[Total: 10]

Cambridge Chemistry Foundation Q6 March 1996 (modified)

Chapter 4

The Periodic Table

The following topics are covered in this chapter:

- *The modern Periodic Table*
- *Redox reactions*
- *The s-block elements: Group 1 and Group 2*
- *Extraction of metals*

4.1 The modern Periodic Table

After studying this section you should be able to:

- *describe the Periodic Table in terms of atomic number, periods and groups*
- *classify the Periodic Table into s, p, d and f blocks*
- *describe and explain periodic trends in atomic radii, ionisation energies, melting and boiling points, and electrical conductivities*

LEARNING SUMMARY

Arranging the elements

In the Periodic Table:

- elements are arranged in order of increasing atomic number
- Groups are vertical columns including elements with similar properties
- Periods are horizontal rows across which there is a trend in properties.

Key areas in the Periodic Table

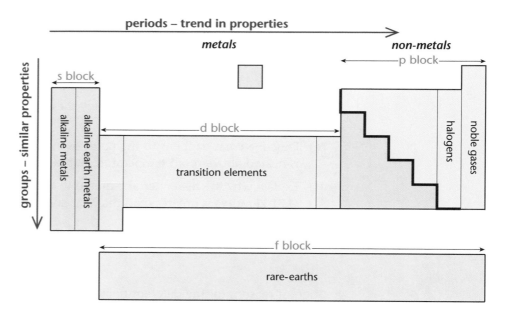

- The diagonal stepped line separates metals (to the left) from non-metals (to the right).
- Elements close to this line, such as silicon and germanium, are called semi-metals or metalloids. They show properties intermediate between those of a metal and a non-metal.
- The four blocks (s, p, d and f) show the sub-shell being filled.

The trend in properties across a period is repeated across each period – this is called *periodicity*.

Periodicity

Periodicity is the periodic trend in properties, repeated across each period.

E.g. Period 2   METAL $\longrightarrow$ NON-METAL

Period 3   METAL $\longrightarrow$ NON-METAL

This periodicity of properties means that predictions can be made about the likely properties of an element and its compounds.

Group 4
C
Si
Ge
Sn
Pb

non-metal to metal

Note, however, the metal/non-metal divide. On descending the Periodic Table, the changeover from metal to non-metal takes place further to the right. For example at the top of Group 4 carbon is a non-metal whereas, at the bottom of Group 4, tin and lead are metals. This means that subtle trends in properties take place down a group.

Trends in atomic radii

AQA	M1	OCR	M1
EDEXCEL	M1	SALTERS	M1
NICCEA	M1	WJEC	CH1
NUFFIELD	M1		

Across a period

Across a period, the atomic radius decreases. The trend in atomic radius across Period 2 is shown below:

Li	Be	B	C	N	O	F
$3p^+$	$4p^+$	$5p^+$	$6p^+$	$7p^+$	$8p^+$	$9p^+$

number of protons increases

radius decreases

Atomic radius decreases across a period.

Across a period:
- the nuclear charge increases
- outer electrons are being added to the same shell
- the attraction between the nucleus and outer electrons increases.

Therefore the atomic radius decreases across a period.

Down a group

Atomic radius increases down a group.

Down a group:
- extra shells are added that are further from the nucleus
- there are more shells between the outer electrons and the nucleus leading to greater shielding of the nuclear charge
- the attraction between the nucleus and the outer electrons decreases.

shells increase

shielding increases

atomic radius increases

Down a group, the nuclear charge also increases but this is more than compensated by the increase in atomic radius and shielding.

Therefore the atomic radius increases down a group.

> Across a Period, nuclear change is the key factor.
>
> Down a Group, atomic radius and shielding are the key factors (see also ionisation energies, page 25).

KEY POINT

Trends in first ionisation energies

AQA	M1	OCR	M1
EDEXCEL	M1	SALTERS	M1
NICCEA	M1	WJEC	CH1
NUFFIELD	M1		

The graph below shows the variation of first ionisation energy with increasing atomic number from hydrogen to calcium:

The graph shows:

> Across a period, increased nuclear charge is most important.

- a general **increase** in first ionisation energy across a period
  (see H→He; Li→Ne; Na→Ar).
  This results from the **increase** in nuclear charge as electrons are added to the same shell across each period.
- a sharp **decrease** in first ionisation energy between the end of one period and the start of the next period (see He→Li; Ne→Na; Ar→K).
  This reflects the addition of a new shell with the resulting increase in **distance** and **shielding**.

> Down a group, increased distance and shielding are most important.

- a **decrease** in first ionisation energy down a group
  (see He→Ne→Ar and other groups).
  This reflects the presence of extra shells down the group.

These trends are discussed in more detail elsewhere in this guide. (See pages 25–27). The reasons for the variation in ionisation energies are similar to those explaining the trend in atomic radii (page 65).

Trends in boiling points

AQA	M1	OCR	M1
EDEXCEL	M1	SALTERS	M1
NICCEA	M1	WJEC	CH1
NUFFIELD	M1		

The graphs below show the variation in boiling points across Period 2 and Period 3 of the Periodic Table.

> Trends in boiling point provide information about structure: giant or simple molecular.

Period 2

> **Giant structure**
> On boiling, strong forces are broken:
> high boiling point.
>
> **Simple molecular structure**
> On boiling, weak forces are broken:
> low boiling point.

Period 3

Across a period:

- the boiling point increases from Group 1 to Group 4
- there is a sharp decrease in boiling point between Group 4 and Group 5
  the boiling points are comparatively low from Group 5 to Group 8.

> **KEY POINT**
>
> The sharp decrease in boiling point marks a change from giant to simple molecular structures.

Note the periodicity in boiling point: the trend for Period 2 is repeated across Period 3.

More details of the structures of these elements are shown below.

| Li | Be | B | C | N_2 | O_2 | F_2 | Ne |
| | giant structures | | | | molecular structures | | |

strong forces between **atoms**   **weak** forces between **molecules**

| Na | Mg | Al | Si | P_4 | S_8 | Cl_2 | Ar |

> Notice the changes in the molecular formulae of P_4, S_8, Cl_2 and Ar. This shows up well in the graph of boiling points across Period 3.
>
> With more atoms (and consequently electrons) in each molecule, the forces between molecules (van der Waals' forces) increase (See also p.55).

Comparison of the boiling points of the metals

Note the increase in boiling points of the metals Na → Al in Period 3.

- The number of delocalised electrons in the lattice increases.
- The charge on each cation increases.
- This results in increasing attractive forces within the metallic lattice.

The diagram below compares the attractive forces between atoms of sodium, magnesium and aluminium.

> See also Metallic bonding, p.49, 60.

Na   Mg   Al

ionic charges increases

number of outer shell electrons increases

attraction increases: boiling point increases

The increasing number of delocalised electrons in the lattice also explains the increase in electrical conductivity from Na → Al.

Progress check

1 Using ideas about nuclear charge, attraction and shells, explain the trend in atomic radii across a period and down each group.
2 Why is the melting point of carbon much higher than that of nitrogen?

1 Period: the nuclear charge increases as electrons are being added to the same shell. The attraction between the nucleus and outer electrons increases; ∴ the atomic radius decreases across a period.
Group: extra shells are added that are further from the nucleus leading to an increased shielding of the outer electrons from the nucleus. Both these factors lead to less attraction between the nucleus and the outer electrons; ∴ the atomic radius increases down a group.
2 Carbon has a giant molecular structure; between the atoms there are strong covalent bonds which break on melting. Nitrogen has a simple molecular structure; between the molecules there are weak van der Waals' forces that break on melting.

4.2 Redox reactions

After studying this section you should be able to:

- *explain the terms reduction and oxidation in terms of electron transfer*
- *explain the terms oxidising agent and reducing agent*
- *apply the rules for assigning oxidation states*
- *identify changes in oxidation numbers from an equation*
- *construct an overall equation for a redox reaction from half-equations*

LEARNING SUMMARY

Oxidation and Reduction

AQA	M2	OCR	M1
EDEXCEL	M1	SALTERS	M2
NICCEA	M1	WJEC	CH1
NUFFIELD	M2		

Redox reactions

Oxidation and *reduction* were originally used for reactions involving oxygen.

> **KEY POINT**
>
> Oxidation is the gain of oxygen.
> Reduction is the loss of oxygen.

Nowadays, oxidation and reduction have a much broader definition in terms of electron transfer in a *redox* reaction (**red**uction and **ox**idation).

OIL

RIG

↓

Oxidation
Is
Loss of electrons
Reduction
Is
Gain of electrons

> **KEY POINT**
>
> **Reduction** is the **gain** of electrons.
> **Oxidation** is the **loss** of electrons.

Reduction and oxidation must take place together:
- if one species **gains** electrons
- another species **loses** the same number of electrons

Half-equations

Half-equations are useful to identify the species being oxidised or reduced. Notice that the number of electrons lost and gained must balance.

The formation of magnesium chloride from its elements is a redox reaction:

overall reaction $Mg + Cl_2 \longrightarrow MgCl_2$

The overall equation conceals the electron transfer that has taken place. This can be shown by writing half-equations:

electron transfer $Mg \longrightarrow Mg^{2+} + 2e^-$ oxidation (loss of electrons)
                        $Cl_2 + 2e^- \longrightarrow 2Cl^-$ reduction (gain of electrons)

Oxidising and reducing agents

Non-metals are oxidising agents.

Metals are reducing agents.

> **KEY POINT**
>
> An **oxidising agent removes** electrons from another reactant.
> Non-metals are oxidising agents, e.g. F_2, Cl_2, O_2.
> A **reducing agent adds** electrons to another reactant.
> Metals are reducing agents, e.g. Na, Fe, Zn.

In the example above:
- Mg is the reducing agent – it has *reduced* the Cl_2 to $2Cl^-$ by *adding* electrons
- Cl_2 is the oxidising agent – it has *oxidised* Mg to Mg^{2+} by *removing* electrons.

Using oxidation numbers

AQA	M2	OCR	M1
EDEXCEL	M1	SALTERS	M2
NICCEA	M1	WJEC	CH1
NUFFIELD	M2		

Chemists use the concept of oxidation number as a means of accounting for electrons.

The oxidation number of a species is assigned by applying a set of rules.

Oxidation number rules

Exceptions to rules

In compounds with fluorine and in peroxides, the oxidation number of oxygen is not –2 and must be calculated from other oxidation numbers.

In metal hydrides, the oxidation number of hydrogen is –1.

species	oxidation number	examples
uncombined element	0	C, zero; Na, zero; O_2, zero.
combined oxygen	–2	H_2O; CaO.
combined hydrogen	+1	NH_3, H_2S.
simple ion	charge on ion	Na^+, +1; Mg^{2+}, +2; Cl^-, –1
combined fluorine	–1	NaF, CaF_2.

> **KEY POINT**
> When applying oxidation numbers to elements, compounds and ions, the sum of the oxidation numbers must equal the overall charge.

Examples of applying oxidation numbers

Oxidation Number rules can be applied to any compound, whether ionic or covalent.

In CO_2, the overall charge is zero.

* There are 2 oxygen atoms, each with an oxidation number of –2, giving a total contribution of –4.
* The oxidation number of carbon must be +4 to give the overall charge of zero.

In NO_3^-, the overall charge is 1–.

* There are 3 oxygen atoms, each with an oxidation number of –2, giving a total contribution of –6.
* The oxidation number of nitrogen must be +5 to give the overall charge of –1.

Oxidation number applies to each atom in a species.

Notice how this has been shown in the examples.

In a compound:
* the total of all the oxidation numbers is zero:

CO_2
+4
–2
–2

+4 –4

CO_2: overall charge = 0   0

In an ion:
* the total of all the oxidation numbers is equal to the overall charge on the ion:

NO_3^-
+5
–2
–2
–2

+5 –6

NO_3^-: overall charge = –1   –1

Using oxidation number with equations

Oxidation numbers can be used to identify redox reactions in which electron loss and electron gain are not easy to see.

By applying oxidation numbers to an equation:

* the species being oxidised and reduced can be identified
* the number of electrons on both sides of the equation can be checked.

The sum of the oxidation numbers on both sides of a chemical equation must be the same.

	Cr_2O_3 (s)	+	2 Al (s)	$\longrightarrow$	Al_2O_3 (s)	+	2 Cr (s)
oxidation	+3 –2		0		+3 –2		0
numbers	+3 –2		0		+3 –2		0
	–2				–2		
sum of oxidation numbers:	0		0		0		0

In this reaction, the changes in oxidation number are:

Cr: $+3 \longrightarrow 0$ reduction   *oxidation number decreases*
Al: $0 \longrightarrow +3$ oxidation   *oxidation number increases*

> **KEY POINT**
>
> Oxidation is an increase in oxidation number.
> Reduction is a decrease in oxidation number.

Combining half-equations

An overall equation for a redox reaction can be constructed by combining the half-equations showing the transfer of electrons in the reaction.

Example: The reaction of Ag⁺ ions react with Zn metal.

electron gain: reduction

electron loss: oxidation

The half-equations are:

$$Ag^+(aq) \; + \; e^- \longrightarrow Ag \, (s)$$
$$Zn(s) \longrightarrow Zn^{2+}(aq) \; + \; 2e^-$$

To combine the half-equations in an overall equation, the number of electrons transferred must be the same in each half-equation. This will ensure that every electron lost by Zn(s) is gained by Ag⁺.

Balance the electrons:
Ag⁺ reaction x 2.

- The Ag⁺ half-equation is multiplied by '2' to balance the electrons.

$$2Ag^+(aq) \; + \; 2e^- \longrightarrow 2Ag(s)$$

- The half-equations are now added:

$$2Ag^+(aq) \; + \; 2e^- \; + \; Zn(s) \longrightarrow 2Ag(s) + Zn^{2+}(aq) + 2e^-$$

Cancel the electrons to give the overall equation.

- Any species appearing on both sides are cancelled to give the overall equation.

$$2Ag^+(aq) \; + \; Zn(s) \longrightarrow 2Ag(s) + Zn^{2+}(aq)$$

- Finally, check that the oxidation numbers balance on either side of the equation:

	$2Ag^+(aq)$	$+$	$Zn(s)$	$\longrightarrow$	$2Ag(s)$	$+$	$Zn^{2+}(aq)$
oxidation numbers:	+1		0		0		+2
	+1				0		

oxidation number check: +2   +2

Progress check

1   What is the oxidation state of the elements in the following:
   (a) Ag^+; (b) F_2; (c) N^{3-}; (d) Fe; (e) MgF_2; (f) $NaClO_3$?

2   Write down the oxidation number of sulphur in the following:
   (a) H_2S; (b) SO_2; (c) SO_4^{2-}; (d) SO_3^{2-}; (e) $Na_2S_2O_3$.

3   The following reaction is a redox process:
   $Mg + 2HCl \longrightarrow MgCl_2 + H_2$
   (a) Identify the changes in oxidation number.
   (b) Which species is being oxidised and which is being reduced?
   (c) Identify the oxidising agent and the reducing agent.

1   (a) +1; (b) 0; (c) −3; (d) 0; (e) Mg: +2, F: −1; (f) Na: +1, Cl: +5, O: −2.
2   (a) −2; (b) +4; (c) +6; (d) +4; (e) +2.
3   (a) Mg, 0 ⟶ +2, H, +1 ⟶ 0; (b) Mg oxidised, H reduced;
    (c) Oxidising agent: HCl, Reducing agent: Mg.

4.3 The s-block elements: Group 1 and Group 2

After studying this section you should be able to:

- describe the characteristic properties of the s-block elements
- understand how flame tests can be used to identify s-block cations
- describe the reactions of the s-block elements with oxygen and with water
- recall some of the reactions of s-block oxides
- recall the relative solubilities of the Group 2 hydroxides and sulphates
- describe and explain the thermal stability of the s-block carbonates and nitrates.
- know that beryllium forms compounds with covalent properties

General properties

AQA	M1	OCR	M1
EDEXCEL	M1	SALTERS	M1
NICCEA	M2	WJEC	CH1
NUFFIELD	M1		

The elements in Group 1 and Group 2 have hydroxides that are alkaline and their common names reflect this.

- The Group 1 elements are the Alkali Metals.
- The Group 2 elements are the Alkaline Earth Metals.

Electronic configuration

The elements in Group 1 and Group 2 have their highest energy electrons in an s sub-shell and these two groups are known as the *s-block elements*.

Each Group 1 element has:
- **one** electron more than the electronic configuration of a noble gas
- an outer s sub-shell containing **one** electron.

Each Group 2 element has:
- **two** electrons more than the electronic configuration of a noble gas
- an outer s sub-shell containing **two** electrons.

> The s-block elements
> Group 1
> Group 2

Electronic configuration of the s-block elements			
Group 1		Group 2	
Li	[He] $2s^1$	Be	[He] $2s^2$
Na	[Ne] $3s^1$	Mg	[Ne] $3s^2$
K	[Ar] $4s^1$	Ca	[Ar] $4s^2$
Rb	[Kr] $5s^1$	Sr	[Kr] $5s^2$
Cs	[Xe] $6s^1$	Ba	[Xe] $6s^2$
Fr	[Rn] $7s^1$	Ra	[Rn] $7s^2$

Physical properties

Group 1
- They are soft metals and can be cut with a knife.
- They have low melting and boiling points.
- They have low densities: Li, Na and K all float on water.
- They have colourless compounds.

Melting points of s-block elements			
Melting point/°C		Melting point/°C	
Li	181	Be	1528
Na	98	Mg	649
K	63	Ca	839
Melting points for comparison: Fe: 1535°C; Cu 1083°C			

Group 2
- They have reasonably high melting and boiling points.
- They have low densities although not as low than those in Group 1.
- They have colourless compounds.

Densities of s-block elements			
Density/g cm^{-3}		Density/g cm^{-3}	
Li	0.53	Be	1.85
Na	0.97	Mg	1.74
K	0.86	Ca	1.54
Densities for comparison: H_2O, 1.00 g cm^{-3}; Fe: 7.86 g cm^{-3}			

Flame tests

EDEXCEL M1   SALTERS M1
NICCEA M1   WJEC CH1
NUFFIELD M1

When heated in a flame, compounds of Groups 1 and 2 show characteristic flame colours, allowing different cations to be identified.

cation	Li^+	Na^+	K^+	Ca^{2+}	Ba^{2+}	Sr^{2+}
flame colour	red	orange	lilac	brick-red	green	red

The characteristic flame colours arise from electronic transitions within the atom.

> Heat energy from the flame excites an electron to a higher energy level.
> The electron returns to a lower energy level.
> The excess energy is released as light energy.
> The frequency of the flame colour depends upon the difference between the energy levels involved.

KEY POINT

Reactivity of the s-block elements

AQA M1   OCR M1
EDEXCEL M1   SALTERS M1
NICCEA M2   WJEC CH1
NUFFIELD M1

The elements in the s-block are the most reactive metals and are strong reducing agents.

Group 1 elements are oxidised in reactions, each atom losing one electron from its outer s sub-shell to form a 1+ ion (+1 oxidation state):

$$M \longrightarrow M^+ + e^-$$

Group 2 elements are oxidised in reactions, each atom losing two electrons from its outer s sub-shell to form a 2+ ion (+2 oxidation state):

$$M \longrightarrow M^{2+} + 2e^-$$

Reactivity **increases** down each group reflecting the increasing ease of losing electrons. The ionisation energy of the metal is an important factor in this process (see also pages 25, 26).

> Within each group, the elements become **more** reactive as the group descends.
> Within each period, the Group 1 element is a **more** reactive metal than the Group 2 element.

KEY POINT

The Group 1 elements

EDEXCEL M1

Reaction with oxygen

The Group 1 elements react vigorously with oxygen. Each element forms the expected ionic oxide with the general formula $(M^+)_2O^{2-}$. However, on descending the group, further oxidation takes place resulting in a mixture of metal oxides.

Lithium forms the expected oxide Li_2O:

$$4Li(s) + O_2(g) \longrightarrow 2Li_2O(s)$$

Sodium forms a mixture of Na_2O and the peroxide Na_2O_2:

When sodium reacts with an excess of oxygen, a peroxide Na_2O_2 is formed.

$$4Na(s) + O_2(g) \longrightarrow 2Na_2O(s)$$
$$2Na(s) + O_2(g) \longrightarrow Na_2O_2(s)$$

Potassium forms a mixture of K_2O, the peroxide K_2O_2 and the superoxide KO_2:

When potassium reacts with an excess of oxygen, a superoxide KO_2 is formed.

$$4K(s) + O_2(g) \longrightarrow 2K_2O(s)$$
$$2K(s) + O_2(g) \longrightarrow K_2O_2(s)$$
$$K(s) + O_2(g) \longrightarrow KO_2(s)$$

reactivity **increases**

Li Be
Na Mg
K Ca
Rb Sr
Cs Ba
Fr Ra

reactivity **increases**

Reaction with water

The Group 1 elements react vigorously with water, forming alkaline solutions of the metal hydroxide and hydrogen.

E.g. $$2Na(s) + 2H_2O(l) \longrightarrow 2NaOH(aq) + H_2(g)$$

The reactivity increases down the group, reflecting the decrease in the first ionisation energies of the elements.

The Group 1 oxides

Reaction with water

The Group 1 oxides react with water forming strongly alkaline solutions.

E.g. $$Na_2O(s) + H_2O(l) \longrightarrow 2NaOH(aq)$$

Reaction with acids

The oxides behave as bases and are neutralised by acids, such as HCl(aq). In each reaction, a salt and water is formed. (See also page 110.)

E.g. $$Na_2O(s) + 2HCl(aq) \longrightarrow 2NaCl(aq) + H_2O(l)$$

The Group 2 elements

AQA	M1	OCR	M1
EDEXCEL	M1	SALTERS	M1
NICCEA	M2	WJEC	CH1
NUFFIELD	M1		

Be
Mg
Ca
Sr
Ba
Ra

reactivity as reducing agent **increases**

Reaction with oxygen

The Group 2 elements react vigorously with oxygen. Each element forms the expected ionic oxide with the general formula, $M^{2+}O^{2-}$.

E.g. $$2Ca(s) + O_2(g) \longrightarrow 2CaO(s)$$

Action of water

Reactivity increases down the group reflecting the increasing ease with which electrons can be lost.

- Mg reacts very slowly with water, forming the hydroxide and hydrogen:
$$Mg(s) + 2H_2O(l) \longrightarrow Mg(OH)_2(aq) + H_2(g)$$

Mg forms MgO with steam.

- With steam, reaction is much quicker forming the oxide and hydrogen:
$$Mg(s) + H_2O(g) \longrightarrow MgO(s) + H_2(g)$$

- Further down the group from calcium, each metal reacts vigorously with water:
$$Ca(s) + 2H_2O(l) \longrightarrow Ca(OH)_2(aq) + H_2(g)$$

Group 2 oxides and hydroxides

$Mg(OH)_2$
$Ca(OH)_2$
$Sr(OH)_2$
$Ba(OH)_2$

solubility increases alkalinity increases

Thus solid barium hydroxide is reasonably soluble in water to form a strong alkaline solution with greater OH⁻(aq) concentration.

Reaction with water

The Group 2 oxides form weakly alkaline solutions with water:

e.g. $$MgO(s) + H_2O(l) \longrightarrow Mg^{2+}(aq) + 2OH^-(aq)$$

- Magnesium hydroxide, $Mg(OH)_2(s)$ is only slightly soluble in water and so the resulting solution is only a weak alkali.

- The solubility in water increases down the group, and resulting solutions are more alkaline:
$$Ba(OH)_2(s) + aq \longrightarrow Ba^{2+}(aq) + 2OH^-(aq)$$

The alkalinity of the Group 2 hydroxides is exploited commercially. Some indigestion tablets contain $Mg(OH)_2$ to neutralise excess acid in the stomach. Farmers add $Ca(OH)_2$ as 'lime' to neutralise acid soils.

MgSO$_4$
CaSO$_4$
SrSO$_4$
BaSO$_4$

solubility **decreases** ⬇

Reaction with acids

- The Group 2 oxides behave as bases and are neutralised by acids, such as HCl(aq), forming salts and water.

    E.g. $MgO(s) + 2HCl(aq) \longrightarrow MgCl_2(aq) + H_2O(l)$

In medicine, barium sulphate is used as a 'barium meal' which shows up imperfections in the gut when exposed to X-rays. Barium compounds are extremely poisonous in solution but the insolubility of BaSO$_4$ is such that no harm is caused to the patient by this treatment.

Solubility of Group 2 sulphates

Magnesium sulphate $MgSO_4$ is very soluble in water but the solubility decreases as the group is descended. The trend is so marked that barium sulphate $BaSO_4$ is virtually insoluble.

In the laboratory, the precipitation of $BaSO_4$ is used to test for the sulphate ion. A solution of a soluble barium salt (usually $BaCl_2(aq)$ or $Ba(NO_3)_2$) is added to a solution of a substance in dilute nitric acid. In the presence of aqueous sulphate ions, a dense white precipitate of barium sulphate is formed.

$$Ba^{2+}(aq) + SO_4^{2-}(aq) \longrightarrow BaSO_4(s)$$

The thermal stability of s-block compounds

EDEXCEL M1 NUFFIELD M1
NICCEA M2

The s-block carbonates

Group 2 carbonates

The Group 2 carbonates are decomposed by heat:

e.g. $MgCO_3(s) \longrightarrow MgO(s) + CO_2(g)$

This is caused by the polarising effect of the Group 2 cation:

MgCO$_3$
CaCO$_3$
SrCO$_3$
BaCO$_3$

ease of thermal decomposition **decreases** ⬇

carbon–oxygen bond is weakened

metal oxide forms

Polarisation occurs when a small cation distorts the electron shells of a larger anion.

The polarising effect is greatest between a small densely charged cation and a large anion.

It is increasingly more difficult to decompose the carbonate as the group is descended. The polarising power of the metal cation decreases as the charge : size ratio decreases (see also page 54).

- As the size of the cation increases, the polarising effect becomes less.
- There is less weakening of the C–O bond.
- It becomes harder to decompose the carbonate.

$[Mg]^{2+}$

$[Ca]^{2+}$

$[Ba]^{2+}$

thermal stability of carbonates increases

increased ionic size gives a less concentrated charge and less polarisation of the carbonate anion

Group 1 carbonates

The Group 1 carbonates are thermally stable (except Li$_2$CO$_3$)

The cations in the Group 1 carbonates are larger than those of the Group 2 carbonates and their polarising power is less. The Li$^+$ ion, however, is comparable in size with the Mg^{2+} ion and lithium carbonate is also decomposed by heat:

$$Li_2CO_3(s) \longrightarrow Li_2O(s) + CO_2(g)$$

The remaining Group 1 carbonates are **not** decomposed by heat.

The s-block nitrates

Group 2 nitrates

The Group 2 nitrates are decomposed by heat, releasing NO_2 and O_2 gases:

e.g. $$Ca(NO_3)_2(s) \longrightarrow CaO(s) + 2NO_2(g) + \tfrac{1}{2}O_2(g)$$

As with the Group 2 carbonates, the nitrates become more thermally stable as the group is descended. This again reflects the increasing size of the cations and the resultant decrease in polarising power.

Group 1 nitrates

The Group 1 nitrates decompose with heat releasing only oxygen gas:

e.g. $$NaNO_3(s) \longrightarrow NaNO_2(s) + \tfrac{1}{2}O_2(g)$$

However, as with the carbonates above, the increasing polarising power of the smaller Li^+ ion makes lithium nitrate less thermally stable and lithium nitrate is decomposed by heat in a similar way to the Group 2 nitrates:

$$2LiNO_3(s) \longrightarrow Li_2O(s) + 2NO_2(g) + \tfrac{1}{2}O_2(g)$$

> With heat, Li_2CO_3 and $LiNO_3$ behave similarly to the carbonates and nitrates of magnesium. This is known as a *diagonal relationship*. You can see why by comparing the positions of Li and Mg in the Periodic Table.

Calcium carbonate

AQA	OCR	M1
EDEXCEL	SALTERS	
NICCEA	WJEC	CH1
NUFFIELD		

Present in limestone and chalk, calcium carbonate is by far the most important calcium compound. It is used in the building industry and in the manufacture of glass and steel.

Much of the calcium carbonate is used in the production of quicklime, CaO.

$$CaCO_3(s) \xrightarrow{\text{heat}} \underset{quicklime}{CaO(s)} + CO_2(g)$$

Addition of water to calcium oxide, produces 'slaked lime', $Ca(OH)_2$.

$$CaO(s) + H_2O(l) \longrightarrow \underset{slaked\ lime}{Ca(OH)_2(s)}$$

In the laboratory, a solution of calcium hydroxide is used as 'lime water' to test for carbon dioxide. The *milky* appearance is a fine suspension of calcium carbonate.

$$\underset{lime\ water}{Ca(OH)_2(aq)} + CO_2(g) \longrightarrow CaCO_3(s) + H_2O(l)$$

If an excess of carbon dioxide is bubbled through lime water, the precipitate disappears with the formation of calcium hydrogencarbonate:

$$CaCO_3(s) + CO_2(g) + H_2O(l) \xrightarrow{excess\ CO_2} Ca^{2+}(aq) + 2HCO_3^-(aq)$$

Calcium hydrogencarbonate is responsible for 'hardness' in water. On being heated, the reverse reaction takes place and calcium carbonate is precipitated as 'scale'.

$$\underset{hard\ water}{Ca^{2+}(aq) + 2HCO_3^-(aq)} \xrightarrow{heat} \underset{scale}{CaCO_3(s)} + CO_2(g) + H_2O(l)$$

> *Quicklime* is used by the building industry to make cement and mortar.
>
> *Slaked lime* is used in mortar and used by farmers to neutralise acid soils.

Progress check

1 Write down equations for the following reactions:
   (a) barium with water
   (b) calcium oxide with nitric acid.

2 Identify the oxidation number changes taking place during the thermal decomposition of sodium nitrate:
   $$2NaNO_3(s) \longrightarrow 2NaNO_2(s) + O_2(g)$$

2 N, $+5 \longrightarrow +3$, O, $-2 \longrightarrow 0$.
1 (a) $Ba(s) + 2H_2O(l) \longrightarrow Ba(OH)_2(aq) + H_2(g)$
  (b) $CaO(s) + 2HNO_3(aq) \longrightarrow Ca(NO_3)_2(aq) + H_2O(l)$

4.4 The Group 7 elements and their compounds

After studying this section you should be able to:

LEARNING SUMMARY

- describe the characteristic properties of the Group 7 elements
- recall the relative reactivity of the halogens as oxidising agents
- recall the characteristic tests for halide ions
- describe the characteristics of hydrogen halides
- describe the reactions of halides with concentrated sulphuric acid
- describe the use of thiosulphate titrations.

General properties

AQA	M2	OCR	M1
EDEXCEL	M1	SALTERS	M2
NICCEA	M1	WJEC	CH1
NUFFIELD	M2		

The common name for the elements in Group 7 is the Halogens.

Electronic configuration

Each halogen has **seven** outer shell electrons, just one electron short of the electronic configuration of a noble gas. The outer **p** sub-shell contains **five** electrons.

Trend in physical states

The halogens exist as diatomic molecules, X_2. The boiling points of the halogens increase in descending the group.

- The physical states of the halogens at r.t.p. show the classic trend of gas $\longrightarrow$ liquid $\longrightarrow$ solid.
- On descending the group the number of electrons increases leading to an increase in the van der Waals' forces between molecules.
- Therefore the boiling point increases.

Electronic configuration of the halogens

F	[He] $2s^2 2p^5$
Cl	[Ne] $3s^2 3p^5$
Br	[Ar] $3d^{10} 4s^2 4p^5$
I	[Kr] $4d^{10} 5s^2 5p^5$
At	[Xe] $4f^{14} 5d^{10} 6p^2 6p^5$

F_2
Cl_2
Br_2
I_2

boiling point **increases** down group

Boiling points of the Halogens

	Boiling point/°C	State at r.t.p.
F_2	−188	gas
Cl_2	−35	gas
Br_2	59	liquid
I_2	184	solid
At_2	337	solid

Trend in electronegativity

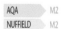

AQA	M2
NUFFIELD	M2

Electronegativity is a measure of the attraction of an atom for the pair of electrons in a covalent bond (see also page 52).

The hydrogen halides have polar molecules: $H^{\delta+}-X^{\delta-}$. The polarity decreases on descending the halogens and the order of polarity is:

most polar   H–F > H–Cl > H–Br > H–I   *least polar*

This trend in polarity results from the **decreasing** electronegativity of the halogen atom on descending the group.

- The atomic radius increases from F → Cl → I resulting in less nuclear attraction on the bonding electrons (this is despite the increase in nuclear charge!).
- There are more electron shells between the nucleus and the bonding electrons to shield the nuclear charge.

electronegativity of halogen **decreases**

polarity of H–X bond **decreases**

Electronegativity of the Halogens

	Pauling value
F	4.0
Cl	3.0
Br	2.8
I	2.5

The overall effect is that the smaller the halogen atom, the greater the nuclear attraction experienced by the bonding electrons. Thus the large electronegativity of fluorine results in a bond between hydrogen and fluorine that is more polar than between hydrogen and other halogens.

The relative reactivity of the halogens as oxidising agents

KEY POINT

The halogens are the most reactive non-metals and are strong oxidising agents.

The halogens become **less** reactive as the group descends as their oxidising power decreases.

The halogens are reduced in reactions, each atom gaining one electron into a p sub-shell to form a –1 ion (–1 oxidation state):

e.g.
$$\tfrac{1}{2}F_2(g) \quad + \quad e^- \quad \longrightarrow \quad F^-(g)$$
$$[\text{He}]\ 2s^2 2p^5 \qquad\qquad\qquad\qquad [\text{He}]\ 2s^2 2p^6\ (\text{or } [\text{Ne}])$$

- An extra electron is captured by being attracted to the outer shell of an atom by the nuclear charge of an atom.

F
Cl
Br
I
At

reactivity as oxidising agent **decreases**

On descending the halogens:

- the atomic radii increase resulting in less nuclear attraction at the edge of the atom (despite the increase in nuclear charge).
- there are more electron shells between the nucleus and the edge of the atom to shield the nuclear charge.

 F

Cl

Br

The overall effect is that most nuclear attraction is experienced at the edge of the small fluorine atoms. This attraction decreases down the group as the atoms get bigger.

Therefore the oxidising power of the halogens decreases down the group.
Fluorine is the strongest oxidising agent and is able to attract an extra electron more strongly than other halogens.

Displacement reactions of the halogens

The decrease in reactivity as the group is descended can be demonstrated by displacement reactions of aqueous halides using Cl_2, Br_2 and I_2.

chlorine oxidises both Br^- and I^-:
$$Cl_2\,(aq)\ +\ 2Br^-\,(aq) \longrightarrow 2Cl^-\,(aq)\ +\ Br_2\,(aq)$$
$$Cl_2\,(aq)\ +\ 2I^-\,(aq) \longrightarrow 2Cl^-\,(aq)\ +\ I_2\,(aq)$$

bromine oxidises I^- only:
$$Br_2\,(aq)\ +\ 2I^-\,(aq) \longrightarrow 2Br^-\,(aq)\ +\ I_2\,(aq)$$

iodine does not oxidise either Cl^- or Br^-

Industrial extraction of bromine

The main source of bromine is as bromide ions, Br^- in sea water. Bromine is extracted by oxidising sea water with chlorine. Because chlorine is a stronger oxidising agent, it displaces the bromide ions using the principles of the displacement reaction.

Testing for halide ions

Addition of aqueous silver ions (using $AgNO_3(aq)$) to aqueous halide ions produces coloured precipitates that have different solubilities in aqueous ammonia.

chloride: $Ag^+(aq) + Cl^-(aq) \longrightarrow AgCl(s)$ white precipitate, soluble in dilute $NH_3(aq)$
bromide: $Ag^+(aq) + Br^-(aq) \longrightarrow AgBr(s)$ cream precipitate, soluble in conc $NH_3(aq)$
iodide: $Ag^+(aq) + I^-(aq) \longrightarrow AgI(s)$ yellow precipitate, insoluble in conc $NH_3(aq)$

Hydrogen halides

| EDEXCEL | M1 | NUFFIELD | M1, M2 |
| NICCEA | M1 | | |

The hydrogen halides, HX, are colourless gases, very soluble in water, forming strongly acidic solutions. For example, hydrochloric acid forms when hydrogen chloride gas dissolves in water:

$$HCl(g) + aq \longrightarrow H^+(aq) + Cl^-(aq)$$

- As the group is descended, the H–X bond enthalpy decreases and the bond breaks more readily, resulting in an increase in acidity down the group.

The order of acidity is HI > HBr > HCl.

- The hydrogen halides are reducing agents with the reducing ability increasing down the group.

Reactions of halides with concentrated sulphuric acid

| AQA | M2 | NICCEA | M1 |
| EDEXCEL | M1 | NUFFIELD | M2 |

HCl
HBr
HI — Increased strength as reducing agents

HCl does **not** reduce H_2SO_4

Concentrated sulphuric acid is an oxidising agent. Halide salts react with concentrated sulphuric acid producing a range of products depending on the halide used. This difference is caused by the increasing reducing power of the hydrogen halides as the group is descended.

NaCl and H_2SO_4

Hydrogen chloride gas, HCl, is formed.

$$NaCl(s) + H_2SO_4(l) \longrightarrow NaHSO_4(s) + HCl(g)$$

The HCl formed is not a sufficiently strong reducing agent to reduce the sulphuric acid. No redox reaction takes place.

NaBr and H_2SO_4

Hydrogen bromide gas, HBr, is initially formed.

$$NaBr(s) + H_2SO_4(l) \longrightarrow NaHSO_4(s) + HBr(g)$$

Some of the hydrogen bromide reduces the sulphuric acid with the formation of sulphur dioxide and orange bromine fumes.

HBr reduces H_2SO_4

$$H_2SO_4 \xrightarrow{HBr} Br_2 + SO_2$$

$$2HBr\,(g) + H_2SO_4(l) \longrightarrow SO_2(g) + Br_2(g) + 2H_2O(l)$$

$$+6 \xrightarrow{\;-2\;} +4 \qquad\qquad \textit{S reduced}$$

$$2 \times -1 \xrightarrow{\;+2\;} 0 \qquad\qquad \textit{Br oxidised}$$

NaI and H_2SO_4

Hydrogen iodide gas, HI, is initially formed.

$$NaI(s) + H_2SO_4(l) \longrightarrow NaHSO_4(s) + HI(g)$$

Hydrogen iodide is a strong reducing agent and reduces the sulphuric acid to a mixture of reduced products including sulphur dioxide and hydrogen sulphide.

Reduction to SO_2 (ox no: +4)

HI reduces H_2SO_4 and SO_2

$$H_2SO_4 \xrightarrow{HI} I_2 + SO_2$$
$$SO_2 \xrightarrow{HI} I_2 + H_2S$$

$$2HI\,(g) + H_2SO_4(l) \longrightarrow SO_2(g) + I_2(s) + 2H_2O(l)$$

$$+6 \xrightarrow{\;-2\;} +4 \qquad\qquad \textit{S reduced}$$

$$2 \times -1 \xrightarrow{\;+2\;} 0 \qquad\qquad \textit{I oxidised}$$

Further reduction to H_2S (ox no: –2)

$$6HI\,(g) + SO_2(g) \longrightarrow H_2S(g) + 3I_2(s) + 2H_2O(l)$$

$$+4 \xrightarrow{\;-6\;} -2 \qquad\qquad \textit{S reduced}$$

$$6 \times -1 \xrightarrow{\;+6\;} 0 \qquad\qquad \textit{I oxidised}$$

Uses of chlorine

AQA	M2	NUFFIELD	M2
EDEXCEL	M1	OCR	M1
NICCEA	M1	WJEC	CH1

> A solution of chlorine in water is green, showing the presence of chlorine.

Small amounts of chlorine are added to drinking water to kill bacteria that would make the water unsafe to drink. Stronger solutions of chlorine in water are in equilibrium (see also page 104):

$$Cl_2(aq) + H_2O(l) \rightleftharpoons HClO(aq) + HCl(aq)$$

HClO chloric(I) acid acts as a bleach and, if indicator solution is added, the initial red colour (from the acids) disappears as the bleaching action takes place.

The formation of bleach

AQA	M2	NUFFIELD	M2
EDEXCEL	M1	OCR	M1
NICCEA	M1		

> Notice how the products obtained depend upon the conditions.
>
> This is an example of how chemists are able to control the products using different conditions.

Bleach is formed when dilute aqueous sodium hydroxide and chlorine are mixed at room temperature.

This reaction is an example of *disproportionation* in which chlorine is both reduced (to chloride, Cl^-) and oxidised (to chlorate(I) ClO^-).

$$Cl_2(aq) + 2\,NaOH(aq) \longrightarrow NaCl(aq) + NaClO(aq) + H_2O(l)$$

$$0 \xrightarrow{\ \ -1\ \ } -1 \qquad \textit{chlorine reduced}$$

$$0 \xrightarrow{\ \ +1\ \ } 1 \qquad \textit{chlorine oxidised}$$

Further disproportionation of the chlorine takes place with hot, concentrated sodium hydroxide:

$$3\,NaClO(aq) \longrightarrow 2\,NaCl(aq) + NaClO_3(aq)$$

$$2 \times +1 \xrightarrow{\ \ -4\ \ } 2 \times -1 \qquad \textit{chlorine reduced}$$

$$+1 \xrightarrow{\ \ +4\ \ } +5 \qquad \textit{chlorine oxidised}$$

The industrial extraction of the chlorine

EDEXCEL	M1

Chlorine

Chlorine is manufactured by electrolysis of a concentrated solution of sodium chloride (purified saturated brine). The diagram below shows the modern 'Membrane cell'.

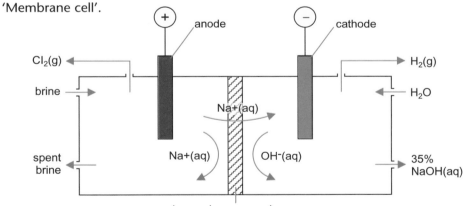

The cell reactions are:

at the cathode $\quad 2H_2O(l) + 2e^- \longrightarrow 2OH^-(aq) + H_2(g)$

at the anode $\quad 2Cl^-(aq) \longrightarrow Cl_2(g) + 2e^-$

> **KEY POINT**
>
> Chlorine reacts with hydroxide ions (see above). The membrane keeps these two products apart, preventing any reaction.
>
> The other products are hydrogen and the alkali, sodium hydroxide.
>
> The process consumes large amounts of electricity, which is the main expense.

Iodine–thiosulphate titrations

NICCEA M1   NUFFIELD M2

Starch is a sensitive test for the presence of iodine. If it is added too early in this titration, the deep-blue colour perseveres and the end-point cannot be detected with accuracy.

Iodine-thiosulphate titrations can be used to determine iodine concentrations, either directly or by liberating iodine using a more powerful oxidising agent.

Iodine oxidises thiosulphate ions:

$$I_2(aq) + 2S_2O_3^{2-}(aq) \longrightarrow 2I^-(aq) + S_4O_6^{2-}(aq)$$

- Thiosulphate is added from the burette to the iodine until the deep-brown colour of iodine becomes a light-straw colour.
- Starch is now added, forming a deep-blue colour.
- Thiosulphate is added dropwise until the deep-blue colour becomes colourless. This is the end-point and shows that all the iodine has just been reacted.

The estimation of chlorine in a bleach

10.0 cm³ of a sample of bleach was diluted to 250 cm³ with water. 25.0 cm³ of this solution was pipetted into a flask. An excess of aqueous potassium iodide was then added to liberate iodine:

equation 1 $Cl_2(aq) + 2I^-(aq) \longrightarrow I_2(aq) + 2Cl^-(aq)$

Compare acid-base titrations (p.40). The essential method is the same.

Titration of this solution required 23.2 cm³ of 0.100 mol dm⁻³ sodium thiosulphate.

equation 2 $2S_2O_3^{2-}(aq) + I_2(aq) \longrightarrow 2I^-(aq) + S_4O_6^{2-}(aq)$

Calculate the concentration of chlorine in the bleach.

From the titration results, amount (in moles) of $Na_2S_2O_3$ can be calculated:

The concentration and volume of $Na_2S_2O_3$ are known.

$$\text{amount (in mol) of } Na_2S_2O_3 = c \times \frac{V}{1000} = 0.100 \times \frac{23.2}{1000} = 0.00232 \text{ mol}$$

From the equations, the amount (in moles) of Cl_2 can be determined:

From *equation 2*, 2 mol $S_2O_3^{2-}$ reacts with 1 mol I_2

$\therefore$ 0.00232 mol $S_2O_3^{2-}$ reacts with $\dfrac{0.00232}{2}$ mol $I_2 = 0.00116$ mol I_2

Using the equations, determine the number of moles of Cl_2.

From *equation 1*, 1 mol Cl_2 forms 1 mol I_2

$\therefore$ amount (in mol) of $Cl_2 = 0.00116$ mol

The concentration in mol dm⁻³ of the diluted bleach solution can now be calculated by scaling to 1000 cm³:

Work out the concentration of $Cl_2(aq)$, in mol dm⁻³.

25.0 cm³ of diluted bleach contains 0.00116 mol Cl_2.

1 cm³ of diluted bleach contains $\dfrac{0.00116}{25}$ mol Cl_2.

1 dm³ of diluted bleach contains $\dfrac{0.00116}{25} \times 1000 = 0.0464$ mol Cl_2.

Finally, take into account the dilution with water from 10 cm³ to 250 cm³:

Don't forget to take into account any dilution.

The original bleach was diluted by a factor of $\dfrac{250}{10} = 25$

$\therefore$ concentration of Cl_2 in the bleach is $25 \times 0.0464 = 1.16$ mol dm⁻³.

Progress check

1 State and explain the trend in boiling points of the halogens fluorine to iodine.

2 How could you distinguish between NaCl, NaBr and NaI by a simple test?

3 Comment on the changes in oxidation number of chlorine in following reaction: $Cl_2(aq) + H_2O(l) \longrightarrow HClO(aq) + H^+(aq) + Cl^-(aq)$

1 Boiling points increase down the group. From F_2 to I_2 the number of electrons increase leading to greater van der Waals' forces between molecules and higher boiling points.

2 Add $AgNO_3(aq)$. NaCl gives a white precipitate, soluble in dilute ammonia. NaBr gives a cream precipitate, soluble in concentrated ammonia. NaI gives a yellow precipitate, insoluble in concentrated ammonia.

3 $Cl_2 \longrightarrow$ HClO, Cl: $0 \longrightarrow +1$ (oxidation). $Cl_2 \longrightarrow Cl^-$, Cl: $0 \longrightarrow -1$ (reduction). Chlorine has been both oxidised and reduced (disproportionation).

4.5 Extraction of metals

After studying this section you should be able to:

- *understand that carbon reduction is used to extract iron from its ore*
- *understand that electrolysis is used to extract aluminium from its ore*
- *understand that metal reduction is used to extract titanium from its ore*
- *understand the importance of economic factors and recycling*

LEARNING SUMMARY

Reduction of metals

AQA   M2

The main methods for extracting metals from their ores are:

- reduction of the ore with carbon
- reduction of the molten ore by the electrolysis
- reduction of the ore with a more reactive metal.

Reduction of metal oxides with carbon

AQA   M2

Production of iron

The main ore of iron, haematite, contains Fe_2O_3 and this is reduced by coke in the blast furnace. This is a continuous process in which high quality haematite, coke and limestone are fed in at the top of the furnace and hot air blown in near the bottom.

> Coke contains a very high carbon content.

> The production of iron in the blast furnace is a **continuous** process. Raw materials are added at the top of the furnace and the products removed at the base. For further production, more raw materials are added and the process continues in a continuous flow, sometimes for months on end.

- Initially, hot coke reacts with air forming carbon monoxide.

$$2C(s) + O_2(g) \longrightarrow 2CO(g)$$

- The carbon monoxide reduces most of the iron oxide at around 1200°C:

$$Fe_2O_3(s) + 3CO(g) \longrightarrow 2Fe(l) + 3CO_2(g)$$

> Both C and CO are reductants in the blast furnace.

- In hotter parts of the furnace, coke also reacts directly with iron oxide:

$$2Fe_2O_3(s) + 3C(s) \longrightarrow 4Fe(l) + 3CO_2(g)$$

This equation represents the overall reaction for the reduction.

Limestone removes SiO_2.

The limestone is used to remove acidic impurities such as silicon dioxide (sand). This produces a slag, largely of calcium silicate, $CaSiO_3$:

$$CaCO_3 + SiO_2 \longrightarrow CaSiO_3 + CO_2$$

- Molten iron collects at the bottom of the furnace and is run off as 'pig iron.'
- Pig iron is very impure and brittle, containing about 4% of carbon (as well as Mn, Si, P and S).

Making steel

Oxygen removes impurities including C, Si, P and Mn.

Steels are alloys of iron with 0.1%–1.0% of carbon. The major process for converting impure iron from a blast furnace into steel is the Basic Oxygen Process.

- Sulphur impurities are first removed by adding magnesium powder to the molten iron. Magnesium sulphide forms in a vigorous reaction and is raked off from the surface of the molten iron.

$$Mg + S \longrightarrow MgS$$

- Oxygen under pressure is blown onto the molten iron.
- Carbon is oxidised to carbon monoxide which escapes as a gas.
- The oxygen also reacts with other impurities (such as Si, P and Mn) in the iron forming a mixture of oxides which remain in the converter.
- Lime is added as a basic oxide and this reacts with any acidic oxides forming a slag which floats to the surface:

e.g. $$CaO + SiO_2 \longrightarrow CaSiO_3$$

Not all of the carbon is removed in the Basic Oxygen Converter.

- 'Mild steel' contains 0.1% carbon.
- 'High carbon steels' contain about 1.0% carbon.

Limitations of carbon reduction

Environmental pollution problems:

sulphide ores form sulphur dioxide;

carbon reduction produces greenhouse gases.

Metal ores often contain metal oxides or sulphides.

- Sulphide ores (such as galena, PbS, and sphalerite, ZnS) are first roasted in air to produce the metal oxide. The sulphur is liberated as sulphur dioxide, which contributes to acid rain.
- Oxide ores (such as haematite, Fe_2O_3) are reduced to the metal using fossil fuels such as coke from coal. Carbon dioxide, emitted as the main gaseous product, is a greenhouse gas.

Ores of some metals (such as titanium and tungsten) react with carbon forming metal carbides, so this is not a practical method for extracting these metals.

$$TiO_2 + 3C \longrightarrow TiC + 2CO$$

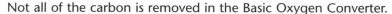

Reduction of metal oxides by electrolysis of the molten ore

AQA   M2
EDEXCEL   M2

For metals more reactive than zinc, reduction with carbon does not take place except at very high temperatures and these metals are usually extracted by electrolysis of the molten ore.

The extraction of aluminium

The melting point of aluminium oxide is 2045°C but this is decreased using molten cryolite. This reduces the energy requirements.

The main ore of aluminium, bauxite, contains Al_2O_3. Purified bauxite is dissolved in molten cryolite (Na_3AlF_6) at 970°C. This is a continuous process needing regular additions of aluminium oxide.

carbon (graphite)   solid crust of electrolyte floats on top

anode (+)

cathode (-)

steel tank lining

carbon (graphite)

molten electrolyte of aluminium oxide dissolved in cryolite   molten aluminium is drawn off at intervals

The electrodes are made of graphite and the cell reactions are:

at the cathode $Al^{3+} + 3e^- \longrightarrow Al$
at the anode $2O^{2-} \longrightarrow O_2 + 4e^-$

- Periodically the graphite anodes need replacing because, at the high temperatures used, the carbon anodes react with the liberated oxygen.
$$2C + O_2 \longrightarrow 2CO$$
$$\text{and } C + O_2 \longrightarrow CO_2$$
- The process consumes large amounts of electricity (as electricity is needed to melt the Al_2O_3 as well as to reduce it).
- The process is only economic where electricity is relatively inexpensive.

Reduction of metal halides with metals

AQA   M2

Some metals, such as titanium, become brittle if contaminated with traces of impurities such as carbon, oxygen or nitrogen.

Highly pure metals can be formed by reduction of a metal halide with a reactive metal.

Despite the high cost, pure titanium is manufactured by reduction of its molten chloride with a more reactive metal.

When pure, titanium is a light metal with a high strength and high resistance to corrosion.

When impure, it is brittle and of little use.

Production of titanium

The main ore of titanium, rutile, contains titanium(IV) oxide. Titanium is manufactured in a two-stage batch process.

- Rutile is first converted to titanium(IV) chloride using chlorine and coke at around 900°C:
$$TiO_2 + 2C + 2Cl_2 \longrightarrow TiCl_4 + 2CO$$
- The titanium(IV) chloride is purified from other chlorides (e.g. those of iron, silicon and chromium) by fractional distillation under argon or nitrogen.
- The chloride is then reduced by a more reactive metal such as sodium or magnesium:
$$TiCl_4 + 4Na \longrightarrow Ti + 4NaCl$$

The production of titanium is a **batch** process. Raw materials are reacted together in a single 'batch' and the products are separated from the reaction mixture. For further production, the whole process needs to be repeated from scratch.

- The sodium is initially held at around 550°C but the temperature rises to nearly 1000°C during the reaction.
- An inert atmosphere of argon is used to prevent any contamination of the metal with oxygen or nitrogen.
- The sodium chloride by-product is washed out with dilute hydrochloric acid, leaving titanium as a granular powder.

Economic factors

- Chlorine and sodium have to be produced first by the electrolysis of molten sodium chloride. This process is expensive.
- High temperatures are involved in both stages of the production.
- Precautions have to be taken in the handling of the intermediate $TiCl_4$ which reacts rapidly with water.
- An argon atmosphere has to be maintained to prevent oxidation of the hot titanium as it forms.

Economic factors and recycling

AQA   M2

Costs of extracting a metal

The method used to reduce a metal on an industrial scale is dependent upon several factors:

The cost of the reducing agent

A reducing agent that is naturally available such as carbon (from coal) is cheaper than one such as sodium, which has to prepared first by a separate (and often costly) process.

The energy costs for the process

The lower the temperature of a process, the lower the energy requirement and the cheaper the process.

The purity required of the metal

It is relatively expensive to produce a metal with very high purity. There needs to be a high demand for the metal if this extra expense is to be justified.

These factors need to be weighed against each other when considering the overall cost of an extraction process.

Recycling

Metals are a valuable resource and, instead of disposal, metals are often collected and recycled.

Recycling of iron

Recycling metals conserves metal ores and saves energy.

- Iron is recycled by collection of scrap iron which is melted and reused.
- The Earth's reserves of iron ore are saved.
- The energy required to mine iron ore, transport it and smelt it is several times greater than the energy required to recycle scrap iron.

Recycling of aluminium

- Owing to its very high resistance to corrosion, used aluminium is as good as new.
- Recycling conserves the Earth's supply of aluminium.
- The electrolysis of aluminium oxide requires high temperatures and vast quantities of electricity.

The cost of re-using scrap aluminium is only one-twentieth of the cost of making the pure metal.

- It is also cheaper and easier to refine recycled aluminium than to mine bauxite and extract the metal by electrolysis.

Sample question and model answer

1

(a) State and explain the trends in atomic radius, electronegativity and boiling point of the halogens from fluorine to iodine.

> Atomic radius increases from fluorine to iodine ✓ because there are more electron shells ✓ and the extra shielding repels the outer electrons. ✓
>
> Electronegativity decreases from fluorine to iodine ✓ because the atomic radius increases ✓ and the attraction for the pair of electrons in a bond decreases. ✓
>
> Boiling point increases from F_2 to I_2 ✓ because the number of electrons in molecules increases. ✓ This results in greater van der Waals' forces between molecules. ✓

[8]

Notice that electronegativity applies to attraction for the electrons in a bond – a point often missed in exams.

Again, you must be careful with language: 'more electrons' results in 'greater van der Waals' forces'.

(b) When chlorine is bubbled into water a pale green solution is formed. If a piece of universal indicator paper is dipped into this solution it first turns red and then becomes white. When chlorine is bubbled into a cold dilute solution of sodium hydroxide a colourless solution is formed.

Explain these observations, write equations for the reactions of chlorine with water and with sodium hydroxide and state why the reaction between chlorine and sodium hydroxide is of commercial importance.

> $Cl_2 + H_2O \longrightarrow HCl + HClO$ ✓
> Acids turn universal indicator red ✓ HClO is a bleach ✓
> $Cl_2 + 2NaOH \longrightarrow NaCl + NaClO + H_2O$ ✓
> NaClO is a commercial bleach/disinfectant ✓

[5]

The reactions of chlorine with water and alkali are often tested in exams.

(c) Explain how aqueous silver nitrate and ammonia solution can be used to show the presence of iodide ions in solution.

> A yellow precipitate forms ✓ which is AgI. ✓ The precipitate is insoluble in concentrated aqueous ammonia. ✓

[3]

[Total: 16]

NEAB Equilibria and Inorganic Chemistry Q8 June 1997

More easy marks for all candidates, allowing credit for learning the chemistry. You should, of course, also know these reactions with chloride and bromide ions.

Practice examination questions

1 In the Periodic Table, describe and explain the trend in the atomic radii of the elements in Period 3 and in Group 1. [Total: 8]

2 Magnesium is in Group II of the Periodic Table.
  (a) Complete, and balance if necessary, the following equations:
   (i) $Mg(s) + H_2O(g) \longrightarrow$
   (ii) $Mg(s) + O_2(g) \longrightarrow$
   (iii) $Mg(s) + H_2O(l) \longrightarrow$ [3]

  (b) Magnesium reacts very slowly with cold water whereas barium reacts rapidly. This shows the increasing reactivity of the Group II elements on descending the group.
   (i)   Suggest why the reactivity increases in this way.
   (ii) Name one other reaction of Group II elements which illustrates this trend of increasing reactivity. [3]

  (c) (i) Magnesium oxide is used as a refractory lining in some furnace linings. What property makes it suitable for its use?
   (ii) Name two other Group II compounds, and state a use for each of them. [3]
  [Total: 9]

  Cambridge Trends and Patterns Q2 March 1997

3
  This question concerns the halogens chlorine, bromine and iodine.
  (a) Describe and explain the trend in oxidising ability shown by the halogens. [5]
  (b) Sodium chlorate(V) can be made by reacting chlorine with hot concentrated aqueous sodium hydroxide.
   $$3Cl_2(g) + 6NaOH(aq) \longrightarrow 5NaCl(aq) + NaClO_3(aq) + 3H_2O(l)$$
   (i)   Use changes in oxidation numbers to show that this is a redox reaction.
   (ii) Calculate the maximum mass of sodium chlorate(V) that could be made from the reaction of 85 dm³ of chlorine at room temperature and pressure.
   (iii) How does chlorine react with dilute aqueous sodium hydroxide at room temperature and what use has the resulting solution? [9]
  [Total: 14]

  Cambridge Trends and Patterns Q7 June 1998 (modified)

4
  (a) (i) Define the term *electronegativity*.
   (ii) State and explain the trend in electronegativity of the halogens fluorine to iodine. [5]

  (b) State and explain the trend in boiling points of the halogens fluorine to iodine. [3]

  (c) Describe what is observed when aqueous bromine is added to separate aqueous solutions of sodium chloride and sodium iodide. Write equation(s) for any reaction(s) occurring. [4]

  (d) State and explain the trend in reducing properties of the halide ions shown by the experiments in part **(c)**. [2]
  [Total: 14]

  NEAB Equilibria and Inorganic Chemistry Q4 June 1995

Chapter 5

Chemical energetics

The following topics are covered in this chapter:

- Enthalpy changes
- Determination of enthalpy changes
- Bond enthalpy

5.1 Enthalpy changes

After studying this section you should be able to:

- understand that reactions can be exothermic or endothermic
- construct a simple enthalpy profile diagram for a reaction
- explain and use the terms: standard conditions, enthalpy changes of reaction, formation and combustion

Energy out, energy in

AQA	M2	OCR	M3
EDEXCEL	M2	SALTERS	M1
NICCEA	M1	WJEC	CH2
NUFFIELD	M1		

(Change in heat content of chemicals)
=
– (Change in heat content of the surroundings)

The first law of thermodynamics underpins all of this section.

Enthalpy, H, is the heat energy that is stored in a chemical system.

Enthalpy cannot be measured experimentally. However, an enthalpy change can be measured from the temperature change in a chemical reaction.

> **KEY POINT**
> An enthalpy change ΔH is the heat energy exchange with the surroundings at constant pressure.

The conservation of energy is an important principle in science. This is often summarised by the first law of thermodynamics.

> **KEY POINT**
> The first law of thermodynamics states that energy may be exchanged between a chemical system and the surroundings but the *total* energy remains constant.

Exothermic reactions

During an exothermic reaction, heat energy is released to the surroundings.

Any energy loss from the chemicals is balanced by the same energy gain to the surroundings, which rise in temperature.

exothermic

chemicals lose energy:
ΔH –ve

surroundings gain energy and rise in temperature:
ΔT +ve

Energy pathway diagram (reaction profile)

Energy loss during reaction

Energy gain by the surroundings (identified by a temperature *rise*: ΔT +ve)

Chemists refer to *surroundings* as anything other than the reacting chemicals.

Surroundings often just means the water in which chemicals are dissolved.

> **KEY POINT**
> In an exothermic reaction, ΔH is negative:
> - heat is given out (to the surroundings)
> - the reacting chemicals lose energy.

Endothermic reactions

During an endothermic reaction, heat energy is taken in from the surroundings.

Any energy gain to the chemicals is provided by the same energy **loss** from the surroundings, which fall in temperature.

Energy pathway diagram (reaction profile)

endothermic

chemicals gain energy:
ΔH +ve

surroundings lose energy and fall in temperature:
ΔT –ve

> In an endothermic reaction, ΔH is positive
> - heat is taken in (from the surroundings)
> - chemicals gain energy.
>
> **KEY POINT**

Standard enthalpy changes

AQA	M2	OCR	M3
EDEXCEL	M2	SALTERS	M1
NICCEA	M1	WJEC	CH2
NUFFIELD	M1, M2		

Enthalpy changes have been measured for many reactions. Many are recorded in data books as *standard enthalpy changes* and these are discussed below.

Standard conditions

$\Delta H^{\ominus}$ refers to an enthalpy (H) change (Δ) under standard conditions ($^{\ominus}$).

Standard pressure is 100 kPa or 1 bar.

The former standard pressure of 101 kPa or 1 atmosphere is still quoted in many books.

> *Standard conditions* are:
> a pressure of 100 kPa
> a stated temperature: 298K (25°C) is usually used
> a concentration of 1 mol dm^{-3} (*for aqueous solutions*).
> A *standard state* is the physical state of a substance under standard conditions.
>
> **KEY POINT**

The standard state of water at 298K and 100 kPa is a liquid.

Standard enthalpy change of reaction

Always use a chemical equation or an unambiguous definition with a stated enthalpy change.

> The *standard enthalpy change of reaction* $\Delta H^{\ominus}_r$ is the enthalpy change that accompanies a reaction in the molar quantities that are expressed in a chemical equation under standard conditions, all reactants and products being in their standard states.
>
> **KEY POINT**

The enthalpy change of reaction $\Delta H^{\ominus}_r$ depends upon the quantities shown in a chemical equation. $\Delta H^{\ominus}_r$ should always be quoted with an equation.

$\Delta H^{\ominus}_r$ only has a meaning with an equation.

$\Delta H^{\ominus}_r$ has units of kJ mol^{-1}

mol^{-1} means 'for the amount (in moles) shown in the equation'.

For the reaction:

$$H_2(g) \quad + \quad \tfrac{1}{2}O_2(g) \longrightarrow H_2O(l) \qquad \Delta H^{\ominus}_r = -286 \text{ kJ mol}^{-1}$$
$$1 \text{ mol} \qquad \tfrac{1}{2} \text{ mol} \qquad 1 \text{ mol}$$

but with twice the quantities, there is twice the enthalpy change:

$$2H_2(g) \quad + \quad O_2(g) \longrightarrow 2H_2O(l) \qquad \Delta H^{\ominus}_r = -572 \text{ kJ mol}^{-1}$$
$$2 \text{ mol} \qquad 1 \text{ mol} \qquad 2 \text{ mol}$$

Standard enthalpy change of combustion

> **KEY POINT**
>
> The *standard enthalpy change of combustion* $\Delta H^{\ominus}_c$ is the enthalpy change that takes place when one mole of a substance reacts completely with oxygen under standard conditions, all reactants and products being in their standard states.

This means that complete combustion of 1 mole of $C_2H_4(g)$ releases 1411 kJ of heat energy to the surroundings at 298 K and 100 kPa.

e.g. $C_2H_4(g) + 3O_2(g) \longrightarrow 2CO_2(g) + 2H_2O(l)$ $\Delta H^{\ominus}_c = -1411$ kJ mol^{-1}

Standard enthalpy change of formation

> **KEY POINT**
>
> The *standard enthalpy change of formation* $\Delta H^{\ominus}_f$ is the enthalpy change that takes place when one mole of a compound in its standard state is formed from its constituent elements in their standard states under standard conditions.

This means that the formation of 1 mole of $H_2O(l)$ from 1 mole of $H_2(g)$ and ½ mole of $O_2(g)$ releases 286 kJ of heat energy to the surroundings at 298K and 100 kPa.

e.g. $H_2(g) + \frac{1}{2}O_2(g) \longrightarrow H_2O(l)$ $\Delta H^{\ominus}_f = -286$ kJ mol^{-1}

For an element, the standard enthalpy change of formation is defined as zero.

The formation of $H_2(g)$ from $H_2(g)$ does not involve a chemical change so there is no enthalpy change.

Standard enthalpy change of neutralisation

Notice that $\Delta H^{\ominus}_f (O_2) = 0$ kJ mol^{-1}
$\Delta H^{\ominus}_r$ for any element must be zero (formation of the element from the element – no change).

> **KEY POINT**
>
> The *standard enthalpy change of neutralisation* $\Delta H^{\ominus}_{neut}$ is the energy change that accompanies the neutralisation of an acid by a base to form 1 mole of $H_2O(l)$, under standard conditions.

$$HCl(aq) + NaOH(aq) \longrightarrow NaCl(aq) + H_2O(l) \quad \Delta H^{\ominus}_{neut} = -57.9 \text{ kJ mol}^{-1}$$
acid   base $\longrightarrow$ 1 mol

- In aqueous solution, strong acids and bases are completely dissociated and $\Delta H^{\ominus}_{neut}$ is approximately equal to the value above. This neutralisation process corresponds to the reaction: $H^+(aq) + OH^-(aq) \longrightarrow H_2O(l)$.
- For weak acids, this enthalpy change is less exothermic because some input of energy is required to dissociate the acid (see also page 109).

Progress check

1 Draw energy profile diagrams for the following reactions:
(a) $N_2O_4(g) \longrightarrow 2NO_2(g)$ $\Delta H = +58$ kJ mol^{-1}
(b) $N_2(g) + 3H_2(g) \longrightarrow 2NH_3(g)$ $\Delta H = -92$ kJ mol^{-1}

2 Write an equation to represent the enthalpy change for:
(a) $\Delta H^{\ominus}_c (CH_4)$;   (b) $\Delta H^{\ominus}_f (NO_2)$

2 (a) $CH_4(g) + 2O_2(g) \longrightarrow CO_2(g) + 2H_2O(l)$
(b) $\frac{1}{2}N_2(g) + O_2(g) \longrightarrow NO_2(g)$

5.2 Determination of enthalpy changes

After studying this section you should be able to:

- calculate enthalpy changes from direct experimental results, using the relationship: energy change = $mc\Delta T$
- use Hess's Law to construct enthalpy cycles
- determine enthalpy changes indirectly, using enthalpy cycles and enthalpy changes of formation and combustion

LEARNING SUMMARY

Direct determination of enthalpy changes

AQA	M2	OCR	M3
EDEXCEL	M2	SALTERS	M1
NICCEA	M1	WJEC	CH2
NUFFIELD	M1, M2		

> The **specific heat capacity** of a substance is the energy required to raise the temperature of 1 g of a substance by 1°C.

Calculating enthalpy changes

The **heat energy change**, Q, in the surroundings can be calculated using the relationship below.

$$Q = mc\Delta T \text{ Joules}$$

- m is the **mass** of the surroundings that experience the temperature change
- c is the **specific heat capacity** of the surroundings
- ΔT is the **temperature change** (final temperature – initial temperature)

Example

Addition of an excess of magnesium to 100 cm³ of 2.00 mol dm⁻³ $CuSO_4(aq)$ raised the temperature from 20.0°C to 65.0°C. Find the enthalpy change for the reaction:

> Assume that solids, such as Mg (s) make little difference to the energy change.

$$Mg(s) + CuSO_4(aq) \longrightarrow MgSO_4(aq) + Cu(s)$$

specific heat capacity of solution c = 4.18 J g⁻¹ K⁻¹;
density of solution = 1.00 g cm⁻³.

Find the heat energy change

100 cm³ of solution has a mass of 100 g;

> Any energy **gain** by the surroundings must have come from the same energy **loss** in the chemical reaction.

Temperature change, ΔT	= (65.0–20.0)°C
	= +45.0°C
Heat energy **gain** to surroundings, $Q = mc\Delta T$	= 100 × 4.18 × 45.0 J
	= +18810 J
∴ heat energy **loss** from the reacting chemicals	= –18810 J

Find out the amount (in moles) that reacted

Amount (in mol) of $CuSO_4$ that reacted = $2.00 \times \dfrac{100}{1000}$ mol = 0.200 mol

Scale the quantities to those in the equation

Mg(s)	+	$CuSO_4(aq)$	$\longrightarrow$	$MgSO_4(aq)$	+	Cu(s)
1 mol		1 mol		1 mol		1 mol

> All values in this example are to 3 significant figures. This indicates the expected accuracy of the answer which should also be expressed to 3 significant figures.
>
> In this case,
> ΔH = –94.1 kJ mol⁻¹.

∴ 0.200 mol (1/5th mol) of $CuSO_4$ $\Rightarrow \Delta H$ of –18810 J

∴ 1 mol of $CuSO_4$ $\Rightarrow \Delta H$ of 5 × –18810 = –94050 J

ΔH = –94.1 kJ mol⁻¹ (to 3 sig. figs.)

∴ enthalpy change of reaction is given by:

$$Mg(s) + CuSO_4(aq) \longrightarrow MgSO_4(aq) + Cu(s) \quad \Delta H = -94.1 \text{ kJ mol}^{-1}$$

Indirect determination of enthalpy changes

AQA	M2	OCR	M3
EDEXCEL	M2	SALTERS	M1
NICCEA	M1	WJEC	CH2
NUFFIELD	M1, M2		

Hess's Law

Many reactions have enthalpy changes that cannot be found directly from a single experiment. Hess's Law provides a method for the *indirect* determination of enthalpy changes.

> Hess's Law is an extension of the First Law of Thermodynamics.

> *Hess's Law* states that, if a reaction can take place by more than one route and the initial and final conditions are the same, the total enthalpy change is the same for each route.

KEY POINT

The diagram below shows two routes for converted reactants into products.

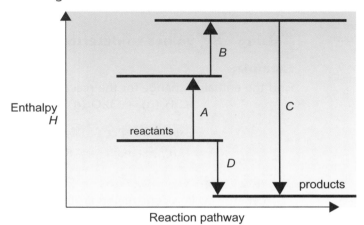

Reaction pathway

> Indirect determination of an enthalpy change uses an energy cycle based on Hess' Law.

> This method is used when the reaction is very difficult to carry out and related reactions can be measured more easily.

Following the arrows from reactants to products,

Route 1:   **A + B + C**
Route 2:   **D**

By Hess's Law, the total enthalpy change is the same for each route.

∴ **A + B + C = D**

If three of these enthalpy changes are known, the fourth can always be calculated.

> Enthalpy changes of combustion are required for all reactants and products.

Using $\Delta H^{\ominus}_c$ values to determine an enthalpy change indirectly

Example:

Find the enthalpy change for the reaction:

$$C(s) + 2H_2(g) \longrightarrow CH_4(g)$$

substance	C(s)	H$_2$(g)	CH$_4$(g)
$\Delta H^{\ominus}_c$ / kJ mol^{-1}	−394	−286	−890

Use the $\Delta H^{\ominus}_c$ data as a 'link' to construct an energy cycle.

- An energy cycle is constructed by linking the reactants and products to their **combustion products**.
- Note the direction of the arrows **from** the reactants and products of the reaction **to** the common combustion products.

> Always show your working. In exams, you are rewarded for a good method.

> If you make one small slip in a calculation, you may only lose 1 mark provided that you show clear working.

> The common combustion products here are CO_2(g) and H_2O(l). You can include these in your cycle but they are not required and 'combustion products' have been used here.

Calculate the unknown enthalpy change
By Hess's Law,

Route 1:	$\Delta H + [(-890)]$
Route 2:	$[(-394) + (2 \times -286)]$

$$\therefore \Delta H + \underbrace{[(-890)]}_{\text{Route 1}} = \underbrace{[(-394) + (2 \times -286)]}_{\text{Route 2}}$$

$\Delta H = [(-394) + (2 \times -286)] - [(-890)]$
$\Delta H = \mathbf{-76 \ kJ \ mol^{-1}}$
$\therefore C(s) + 2H_2(g) \longrightarrow CH_4(g) \qquad \Delta H^{\ominus} = -76 \ kJ \ mol^{-1}$

> **KEY POINT**
>
> **For enthalpy changes of combustion data only,**
>
> $\Delta H = \Sigma \Delta H^{\ominus}_c \text{ (reactants)} - \Sigma \Delta H^{\ominus}_c \text{ (products)}$

Using $\Delta H^{\ominus}_f$ values to determine an enthalpy change indirectly

> Enthalpy changes of formation are required for all reactants and products that are compounds.
>
> For elements, $\Delta H^{\ominus}_f = 0$ (formation of the element from the element – no change).

Example:
Find the enthalpy change for the reaction:

$$C_2H_6(g) + 3\tfrac{1}{2}O_2(g) \longrightarrow 2CO_2(g) + 3H_2O(l)$$

substance	C_2H_6 (g)	CO_2 (g)	H_2O (l)
$\Delta H^{\ominus}_f$/ kJ mol^{-1}	−85	−394	−286

Use the $\Delta H^{\ominus}_f$ data as a 'link' to construct an energy cycle
- An energy cycle is constructed linking the reactants and products with their constituent elements.

> Notice that
> $\Delta H^{\ominus}(O_2) = 0 \ kJ \ mol^{-1}$
> This can be omitted from the cycle.

- Note the direction of the arrows from the constituent elements to the reactants and products of the reaction.

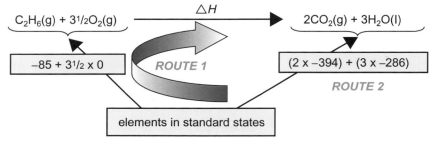

Calculate the unknown enthalpy change
By Hess's Law,

Route 1:	$[(-85) + (3\tfrac{1}{2} \times 0)] + \Delta H$
Route 2:	$[(2 \times -394) + (3 \times -286)]$

$$\therefore \underbrace{[(-85) + 0] + \Delta H}_{\text{Route 1}} = \underbrace{[(2 \times -394) + (3 \times -286)]}_{\text{Route 2}}$$

$\therefore \Delta H = [(2 \times -394) + (3 \times -286)] - [(-85) + 0]$
$\therefore \Delta H = \mathbf{-1561 \ kJ \ mol^{-1}}$
$C_2H_6(g) + 3\tfrac{1}{2}O_2(g) \longrightarrow 2CO_2(g) + 3H_2O(l) \qquad \Delta H^{\ominus} = -1561 \ kJ \ mol^{-1}$

> **KEY POINT**
>
> **For enthalpy changes of formation data only,**
>
> $\Delta H = \Sigma \Delta H^{\ominus}_f \text{ (products)} - \Sigma \Delta H^{\ominus}_f \text{ (reactants)}$

Progress check

For questions 1 and 2 assume that:

specific heat capacity of solution
$c = 4.18 \text{ J g}^{-1} \text{ K}^{-1}$;

density of solution
$= 1.00 \text{ g cm}^{-3}$.

1 Addition of zinc powder to 55.0 cm³ of aqueous copper(II) sulphate at 22.8 °C raised the temperature to 32.3 °C. 0.3175 g of copper were obtained.
(a) Calculate the energy released in this reaction.
(b) Write an equation, including state symbols, for this reaction.
(c) Calculate the enthalpy change for this reaction per mole of copper formed.

2 Combustion of 1.60 g of ethanol, C_2H_5OH, raised the temperature of 150 g of water from 22.0 °C to 71.0 °C. Find the enthalpy change of combustion of ethanol.

3 Use the $\Delta H^{\ominus}_c$ data below to calculate enthalpy changes for:
(a) $3C(s) + 4H_2(g) \longrightarrow C_3H_8(g)$
(b) $C(s) + 2H_2(g) + \tfrac{1}{2}O_2(g) \longrightarrow CH_3OH(l)$

substance	$\Delta H^{\ominus}_c$ / kJ mol⁻¹
C(s)	−394
$H_2(g)$	−286
$C_3H_8(g)$	−2219
$CH_3OH(l)$	−726

4 Use the $\Delta H^{\ominus}_f$ data below to calculate enthalpy changes for:
(a) $C_2H_4(g) + H_2(g) \longrightarrow C_2H_6(g)$
(b) $SO_2(g) + 2H_2S(g) \longrightarrow 2H_2O(l) + 3S(s)$

compound	$\Delta H^{\ominus}_f$ / kJ mol⁻¹
$C_2H_4(g)$	+52
$C_2H_6(g)$	−85
$SO_2(g)$	−297
$H_2S(g)$	−21
$H_2O(l)$	−286

4 (a) −137 kJ mol⁻¹; (b) −233 kJ mol⁻¹.
3 (a) −107 kJ mol⁻¹; (b) −240 kJ mol⁻¹.
2 −883 kJ mol⁻¹.
1 (a) 2.18 kJ; (b) $Zn(s) + CuSO_4(aq) \longrightarrow Cu(s) + ZnSO_4(aq)$; (c) −437 kJ mol⁻¹.

5.3 Bond enthalpy

After studying this section you should be able to:

- *understand and use the term bond enthalpy*
- *explain chemical reactions in terms of enthalpy changes associated with the breaking and making of chemical bonds*
- *determine enthalpy changes indirectly, using average bond enthalpies*

LEARNING SUMMARY

Bond enthalpy

AQA	M2	OCR	M3
EDEXCEL	M2	SALTERS	M1
NICCEA	M1	WJEC	CH2
NUFFIELD	M2		

Enthalpy is stored within chemical bonds and bond enthalpy indicates the strength of a chemical bond in a gaseous molecule. For simple molecules, such as $H_2(g)$ and $HCl(g)$, bond enthalpy applies to the following processes:

$$H-H(g) \longrightarrow 2H(g) \qquad \Delta H = +436 \text{ kJ mol}^{-1}$$
$$H-Cl(g) \longrightarrow H(g) + Cl(g) \quad \Delta H = +432 \text{ kJ mol}^{-1}$$

> Bond enthalpies are **positive** and refer to **bond breaking** – this process requires energy.

> **Bond enthalpy** is the enthalpy change required to break and separate 1 mole of bonds in the molecules of a gaseous element or compound so that the resulting gaseous species exert no forces upon each other.

KEY POINT

> See also Activation Energy: p.99.

Average bond enthalpies

Bond enthalpies such as those above (H–H and H–Cl) apply to specific compounds. Only H_2 can have H–H bonds and the H–H bond enthalpy shown above has a definite value. However, some bonds can have different strengths in different environments.

> Not all C–H bonds are created equal.

The Cl atom in CH_3Cl affects the environment of the C–H bonds.

C—H bonds have different strengths and different bond enthalpies

Data books provide an indication of the likely bond enthalpy of a particular bond by listing average or mean bond enthalpies.

> An *average* bond enthalpy indicates the strength of a *typical* bond. The average bond enthalpy for the C–H bond is +413 kJ mol^{-1}.

> Bond breaking requires energy: ENDOTHERMIC .
>
> Bond making releases energy: EXOTHERMIC.

Bond making and bond breaking

Chemical reactions involve bond breaking followed by bond making.

- Energy is first needed to break bonds in the reactants.
  Bond breaking is an endothermic process and **requires** energy.
- Energy is then released as new bonds are formed in the products.
  Bond making is an exothermic process and **releases** energy.

> Bond enthalpy is an endothermic change (ΔH +ve) for bonds being broken.
>
> When bonds are made, the enthalpy change will be the same magnitude but the opposite sign (ΔH –ve).

Using bond enthalpies to determine enthalpy changes

The enthalpy change for a reaction involving simple gaseous molecules can be determined using average bond enthalpies in an energy cycle:

- Enthalpy required to break bonds = Σ(bond enthalpies of reactants)
- Enthalpy released to make bonds = $-\Sigma$(bond enthalpies of products)

Notice that the relative strengths of the bonds in the reactants and the bonds in the products decide whether a reaction is exothermic or endothermic.

The enthalpy change for a reaction involving gaseous molecules can be conveniently calculated using **average** bond enthalpies.

The calculated value however is only approximate – the **actual** bond enthalpies involved may differ from the average values.

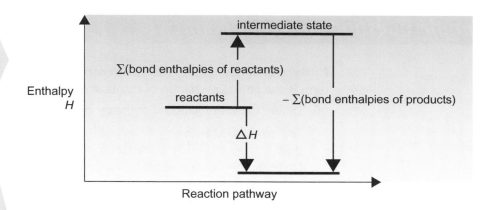

Enthalpy *H*

Reaction pathway

From the diagram above,
$\Delta H = \Sigma$(bond enthalpies of reactants) – Σ(bond enthalpies of products).

For the reaction: $CH_4(g)$ + $2O_2(g)$ ⟶ $CO_2(g)$ + $2H_2O(g)$,

$$4(C–H) + 2(O=O) \longrightarrow 2(C=O) + 4(O–H)$$

ΔH/kJ mol^{-1} $(4 \times 413) + (2 \times 497)$ $(2 \times 740) + (4 \times 463)$

Bonds broken: (endothermic)   *Bonds made: (exothermic)*

$\Delta H = \Sigma$(bond enthalpies of reactants) – Σ(bond enthalpies of products)
$\therefore \Delta H = [(4 \times 413) + (2 \times 497)] - [(2 \times 740) + (4 \times 463)]$ kJ mol^{-1}
= **–686 kJ mol^{-1}**

Progress check

1 You are provided with data for the following reaction:
   $H_2(g) + I_2(g) \longrightarrow 2HI(g)$ $\Delta H = +53$ kJ mol^{-1}
   $H_2(g) + I_2(g) \longrightarrow 2H(g) + 2I(g)$ $\Delta H = +183$ kJ mol^{-1}
   (a) Draw an energy profile diagram for this reaction showing both these energy changes.
   (b) Use your diagram to find out how much energy is released during bond formation in this reaction.
   (c) In this reaction, what can be deduced about the relative strengths of the bonds that are broken and the bonds that are formed?

2 Use the data in the margin to find the enthalpy change for each reaction.
   (a) $C_2H_4(g) + 3O_2(g) \longrightarrow 2CO_2(g) + 2H_2O(g)$
   (b) $N_2(g) + 3H_2(g) \longrightarrow 2NH_3(g)$

bond	bond average enthalpy /kJmol^{-1}
C–H	+413
C=O	+740
O=O	+497
O–H	+463
C=C	+612
C–C	+347
N≡N	+945
H–H	+436
N–H	+391

2 (a) –1057 kJ mol^{-1}; (b) –93 kJ mol^{-1}.

(c) Bonds broken are stronger than bonds that are made.

1 (a), (b)

Sample question and model answer

Enthalpy changes enable chemists to compare different fuels. A particularly useful term is the enthalpy change of combustion.

(a) Define the term standard enthalpy change of combustion.

The standard enthalpy change of combustion is the enthalpy change that accompanies the complete combustion ✓ of one mole ✓ of a compound under standard conditions. [2]

> The question uses 3 significant figures and your answer should also use this accuracy.

(b) In an experiment, the combustion of 0.00175 mol propan-1-ol C_3H_7OH produced sufficient heat to raise the temperature of 50.0 g of water by 15.0°C.

(i) Calculate the energy released by the combustion of 0.00175 mol propan-1-ol. [Specific heat capacity, c, of water = 4.18 J g⁻¹ K⁻¹.]

energy released by combustion of propan-1-ol = energy gained by water
energy gained by water = mass water × specific heat capacity × temperature change
= 50.0 × 4.18 × 15.0 J = 3135 J
= 3.14 kJ (to 3 sig. figs.) ✓

(ii) Calculate the experimental enthalpy change of combustion of propan-1-ol.

> Remember to include the sign! This reaction is exothermic and the negative sign is essential.

enthalpy change from combustion of 0.00175 mol propan-1-ol = −3.14 kJ
The definition for enthalpy change of combustion refers to 1 mole.
∴ $\Delta H_{combustion}$ = −3.14/.00175 ✓ = −1790 kJ mol⁻¹ ✓ (to 3 sig. figs.) [3]

(c) Using the data below, show that the theoretical enthalpy change of combustion of propan-1-ol is −2022 kJ mol⁻¹.

substance	$C_3H_7OH(l)$	$CO_2(g)$	$H_2O(l)$
$\Delta H^{\ominus}_f$ / kJ mol⁻¹	−304	−394	−286

The equation for this enthalpy change is:
$C_3H_7OH(l) + 4\frac{1}{2}O_2(g) \longrightarrow 3CO_2(g) + 4H_2O(l)$ ✓
Linking this equation to the ΔH data gives an enthalpy cycle:

> The correct cycle is critical. Make sure the arrows go in the direction indicated by the definitions.
>
> Here the data are ΔH values and the link is **from** the constituent elements **to** the reactants and products.

$C_3H_7OH(l) + 4\frac{1}{2}O_2(g)$ ——ΔH——→ $3CO_2(g) + 4H_2O(l)$ ✓ (cycle)

−304 kJ mol⁻¹   (3 × −394) + (4 × −286) kJ mol⁻¹

$3C(s) + 4H_2 + 5O_2(g)$

From the cycle, (−304) + ΔH = (3 × −394) + (4 × −286) ✓
∴ ΔH = −2022 ✓ kJ mol⁻¹ [4]

(d) This experiment was repeated several times to obtain an average 'experimental' value of −1800 kJ mol⁻¹. Compare this average 'experimental' value with the theoretical value in **(c)** and suggest **two** reasons for any difference.

> The word 'suggest' indicates that there may not be a 'correct answer'. Any **sensible** response would be credited.
>
> DON'T blame shoddy practical techniques – these are **your** fault.

non-standard conditions/ heat loss/ evaporation of propanol/ incomplete combustion (up to 2 valid points) ✓ ✓ [2]

[Total: 11]

Cambridge Chemistry Foundation Q5 June 1996

Practice examination questions

1 (a) Write a chemical equation, including state symbols, for the reaction which is used to define the enthalpy of formation of one mole of aluminium chloride ($AlCl_3$). [2]

(b) State the additional condition necessary if the enthalpy change for this reaction is to be the standard enthalpy of formation, $\Delta H^{\ominus}_f$, at 298K. [1]

(c) Use the standard enthalpies of formation given below to calculate a value for the standard enthalpy change for the following reaction:

$$AlCl_3(s) + 6H_2O(l) \longrightarrow AlCl_3.6H_2O(s)$$

substance	$AlCl_3(s)$	$H_2O(l)$	$AlCl_3.6H_2O(s)$
$\Delta H^{\ominus}_f$ / kJ mol^{-1}	695	286	2680

[3]
[Total: 6]

NEAB Atomic Structure, Bonding and Periodicity Q5 Feb 1996 (modified)

2 Methanol is used as a fuel making use of combustion reaction with oxygen.

(a) (i) Define the term *standard enthalpy change of combustion*.
  (ii) State the temperature that is conventionally chosen for quoting standard enthalpy changes. [3]

(b) Calculate the standard enthalpy change of combustion of methanol using the following data.

compound	$\Delta H^{\ominus}_f$ /kJ mol^{-1}
$CH_3OH(l)$	−239
$CO_2(g)$	−394
$H_2O(l)$	−286

[3]

(c) The combustion of methanol is a highly exothermic reaction, but methanol does not spontaneously burst into flame and requires a spark for ignition. Explain why this should be so. [1]

[Total: 7]

Cambridge Chemistry Foundation Q3 June 1997 (modified)

3 (a) Define the term *standard enthalpy of formation* ($\Delta H^{\ominus}_f$). [3]

(b) Give the chemical equation for which the enthalpy change is the standard enthalpy of formation of methane. [2]

(c) Use the following data to calculate a value for the enthalpy of formation of methane.

$$C(s) + O_2(g) \longrightarrow CO_2(g) \qquad \Delta H = -394 \text{ kJ mol}^{-1}$$
$$H_2(g) + \tfrac{1}{2}O_2(g) \longrightarrow H_2O(g) \qquad \Delta H = -242 \text{ kJ mol}^{-1}$$
$$CH_4(g) + 2O_2(g) \longrightarrow CO_2(g) + 2H_2O(g) \qquad \Delta H = -802 \text{ kJ mol}^{-1}$$

[3]

(d) Use the following data to calculate mean bond enthalpy values for the C–H and the C–C bonds.

$$CH_4(g) \longrightarrow C(g) + 4H(g) \qquad \Delta H = 1648 \text{ kJ mol}^{-1}$$
$$C_2H_6(g) \longrightarrow 2C(g) + 6H(g) \qquad \Delta H = 2820 \text{ kJ mol}^{-1}$$

[3]

[Total: 11]

NEAB Atomic Structure, Bonding and Periodicity Q4 June 1995

Chapter 6

Rates and equilibrium

The following topics are covered in this chapter:

- *Reaction rates*
- *Catalysis*

- *Chemical equilibrium*
- *Acids and bases*

6.1 Reaction rates

> **LEARNING SUMMARY**
>
> *After studying this section you should be able to:*
> - *recall the factors which control the rate of a chemical reaction*
> - *understand, in terms of collision theory, the effect of concentration changes on the rate of a reaction*
> - *explain the significance of activation energy using the Boltzmann distribution*
> - *use the Boltzmann distribution to explain the effect of temperature changes on a reaction rate*
> - *understand that an enthalpy change indicates the feasibility of a reaction*

What is a reaction rate?

AQA	M2	OCR	M3
EDEXCEL	M2	SALTERS	M1, M2
NICCEA	M2	WJEC	CH2
NUFFIELD	M2		

The rate of a chemical reaction is a measure of how quickly a reaction takes place.

> **KEY POINT**
>
> The rate of a reaction is usually measured as the rate of change of concentration of a stated species in a reaction.
>
> The units of rate are mol dm^{-3} s^{-1}.

Measuring reaction rates

Although concentration can be measured, other quantities proportional to concentration, such as gas volumes, may be easier to monitor. Rates of different reactions show a very wide variation and it may be more convenient to measure a reaction rate per minute or per hour rather than per second.

What affects a reaction rate?

The reasons why reaction rates are affected by these factors is discussed in the rest of this chapter.

The rate of a reaction may be affected by the following factors:

- the concentration of the reactants
- the surface area of solid reactants
- a temperature change
- the presence of a catalyst.

In some reactions, such as photosynthesis, the rate is affected by the presence and intensity of radiation.

How does reaction rate change during a reaction?

A is being used up – its concentration decreases.

B is being formed – its concentration increases.

For a reaction: **A** $\longrightarrow$ **B**, the reaction rate = $\dfrac{\text{change of concentration}}{\text{time}}$
The reaction rate can be measured as:

- the rate of decrease in concentration of **A**
- the rate of increase in concentration of **B**.

The graphs below show how the concentrations of **A** and **B** change during the course of the reaction: **A** ⟶ **B**.

A is used up and its concentration falls. **B** is formed and its concentration increases.

> In the graph, the gradient indicates the reaction rate at any time.

- The steeper the gradient, the faster the rate of the reaction.
- The reaction is fastest at the start when the concentration of **A** is greatest.
- As the reaction proceeds, the rate slows down because the concentration of **A** decreases.
- When the reaction is complete, the graph levels off and the gradient becomes zero.

Activation energy

AQA	M2	OCR	M3
EDEXCEL	M2	SALTERS	M2
NICCEA	M2	WJEC	CH2
NUFFIELD	M2		

> The activation energy of a reaction is the minimum energy required for the reaction to occur.

In a gas or solution, particles are in constant motion and they collide both with each other, with any solid species and with the walls of their container. When particles collide, a reaction can only take place if the energy of the collision exceeds the activation energy of the reaction.

The *activation energy* of a reaction is the energy required to start a reaction by breaking bonds (see page 94).

> Only those collisions of sufficient energy to overcome the activation energy lead to a reaction.

- Activation energy is often supplied by a spark or by heating the reactants.
- Reactions with a small activation energy often take place very readily.
- A large activation energy may 'protect' the reactants from taking part in the reaction (at room temperature).

The Boltzmann distribution

AQA	M2	OCR	M3
EDEXCEL	M2	SALTERS	M2
NICCEA	M2	WJEC	CH2
NUFFIELD	M2		

The Boltzmann distribution shows the distribution of molecular energies in a gas at constant temperature.

The Boltzmann distribution

Characteristics of the Boltzmann distribution

- Most gas molecules have energies within a comparatively narrow range.
- The curve will only meet the energy axis at infinity energy. No molecules have zero energy.
- The area under the distribution curve gives the total number of gas molecules.
- Only those molecules with more energy than the activation energy of the reaction are able to react.

The effect of a concentration change on reaction rate

If the concentration of a reactant in solution or in a gas mixture is increased,

A reaction can only take place if the activation energy is exceeded.

- there are more particles present per volume
- more collisions take place each second
- more collisions exceed the activation energy every second
- therefore the rate of reaction increases.

The diagram below shows distribution curves for two concentrations, C^1 and C^2. where concentration, C_2 > concentration, C_1.

- Providing the temperature is the **same**, distribution curves for different concentrations have the **same** shape.

Note that the proportion of the total number of molecules exceeding the activation energy is the same. The rate increases because there are more molecules per volume and more molecules must now exceed the activation energy.

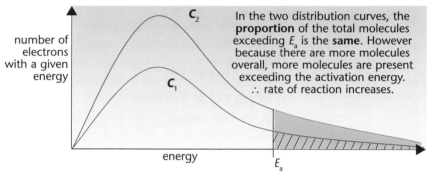

number of electrons with a given energy

In the two distribution curves, the **proportion** of the total molecules exceeding E_a is the **same**. However because there are more molecules overall, more molecules are present exceeding the activation energy.
∴ rate of reaction increases.

energy

E_a

For reactions involving gases, increasing the pressure also increases the concentration of any gas. This results in an increased reaction rate.

The effect of a change in surface area on reaction rate

For a reaction involving a solid, the reaction takes place at a faster rate when the solid is in a powdered form rather than as lumps.
Calcium carbonate reacts with hydrochloric acid producing carbon dioxide gas:

$$CaCO_3(s) + 2HCl(aq) \longrightarrow CaCl_2(aq) + H_2O(l) + CO_2(g)$$

By measuring the volume of carbon dioxide gas evolved with time, the rate of this reaction can be monitored. The graph below compares the reaction rates when using an excess of calcium carbonate as powder and as lumps. The same volume of the same concentration of hydrochloric acid has been reacted in both experiments.

The gradient provides an indication of reaction rate.

The steeper the curve, the faster the reaction.

The final volume is the same in both experiments since the same amount (in mol) of hydrochloric acid has been reacted with an excess of calcium carbonate.

volume of CO_2

powdered $CaCO_3$

lumps of $CaCO_3$

time

The gradient of the graph is much steeper with powdered carbonate because the surface area available for the reaction with hydrochloric acid is much greater than with lumps allowing more collisions per second.

The effect of a temperature change on reaction rate

The average kinetic energy of the particles is proportional to temperature. As temperature increases, so does the kinetic energy of gas molecules.

The diagram below shows distribution curves for a sample of gas at two temperatures, T_1 and T_2, where temperature, $T_2 >$ temperature, T_1.

Increasing the temperature **moves** the distribution curve to the right.

Increasing the temperature **does not change** the activation energy or the total number of molecules – the **shape of the curve changes**.

A **greater proportion** of molecules **exceeds the activation energy at higher temperature**.

The Boltzmann distribution curve is displaced to the right with the peak lower. The average energy is now increased.

The total area under each curve is a measure of the total number of molecules present, and this is the same for each curve.

An increase in temperature increases the rate of a reaction because:

- the molecules move faster and have more kinetic energy
- there are more collisions each second
- the increased kinetic energy produces more energetic collisions
- a greater proportion of molecules exceed the activation energy.

The feasibility of a chemical reaction

EDEXCEL	M2
NICCEA	M2
WJEC	CH2

The value of the enthalpy change for a reaction provides an indication of the *relative stability* of reactants and products and the *feasibility* of a reaction.

For example, sucrose is oxidised by oxygen in a highly exothermic reaction:

$$C_{12}H_{22}O_{11}(s) + 12O_2(g) \longrightarrow 12CO_2(l) + 11H_2O(g) \quad \Delta H = -5644 \text{ kJmol}^{-1}$$

- The reaction is *very likely* to take place because of the great amount of energy released in this reaction – the products are *thermodynamically* far more stable.

Activation energy is the energy barrier of a reaction.

- However, the reaction does not take place spontaneously at room temperature because the reaction has a large activation energy which acts as an energy barrier – the reactants are *kinetically* stable.

Progress check

1 Explain the following, in terms of collision theory and distribution graphs.
   (a) Reactions take place quicker when the reactants are more concentrated.
   (b) Reactions take place quicker at higher temperatures.

1 (a) Because there are more molecules per volume, more collisions take place each second. More collisions exceed the activation energy every second, increasing the reaction rate.
(b) The molecules move faster and have more kinetic energy. There are more collisions each second and the increased kinetic energy produces more energetic collisions. A greater proportion of molecules exceeds the activation energy.

6.2 Catalysis

After studying this section you should be able to:

- *understand what is meant by a catalyst*
- *explain how a catalyst changes the activation energy of a reaction*
- *explain what is meant by homogeneous and heterogeneous catalysis*

LEARNING SUMMARY

How do catalysts work?

AQA	M2	OCR	M1
EDEXCEL	M2	SALTERS	M1, M2
NICCEA	M2	WJEC	CH2
NUFFIELD	M2		

What is a catalyst?

A catalyst changes the rate of a chemical reaction, but is unchanged at the end of the reaction. Most catalysts speed up reaction rates although there are some (called inhibitors) that slow down reactions.

These key points are often tested in exams.

> **KEY POINT**
> A catalyst speeds up the rate of a reaction by providing an **alternative route** for the reaction with a **lower activation energy**.
> Activation energy with catalyst, E_c < Activation energy without catalyst, E_a

The effects of a catalyst on reaction route and activation energy are shown in the diagrams below.

Energy profile diagram

Notice the way the catalyst reduces the energy barrier to the reaction.

The catalyst provides an alternative route in which $E_c < E_a$

The catalyst does **not** change the distribution curve.

Notice how more molecules have an energy exceeding the new, lower activation energy.

Boltzmann distribution curve

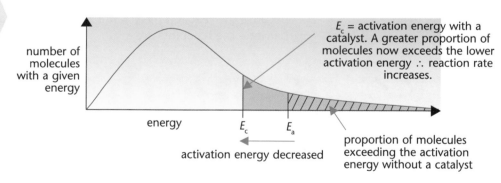

E_c = activation energy with a catalyst. A greater proportion of molecules now exceeds the lower activation energy ∴ reaction rate increases.

activation energy decreased

proportion of molecules exceeding the activation energy without a catalyst

Heterogeneous and homogeneous catalysis

OCR	M3	NICCEA	M2
SALTERS	M1, M2	WJEC	CH2

A **heterogeneous** catalyst is in a **different** phase from the reactants. Many reactions between **gases** are catalysed using a **solid** catalyst.

hetero- = different
homo- = same

A **homogeneous** catalyst is in the **same** phase as the reactants. Many reactions between **gases** in the atmosphere are catalysed using a **gaseous** catalyst. Homogeneous catalysis also takes place in many **aqueous** reactions in which both the reactants and the catalyst are dissolved in water.

Many heterogeneous catalysts are transition metals or their compounds.

Heterogeneous catalysis

The examples below are gaseous reactions catalysed by a solid catalyst.

Haber Process for ammonia (see also pages 107–108)

$$N_2(g) + 3H_2(g) \rightleftharpoons 2NH_3(g)$$

- The gaseous equilibrium process is catalysed by a solid iron catalyst, Fe(s).

surface of iron catalyst
Reactant molecules diffuse towards the catalyst surface

Reactant molecules adsorbed at active sites on the catalyst surface

Bonds weaken allowing reaction to take place

Following reaction, the product molecules diffuse away from the catalyst surface

Polluting gases temporarily bond to the catalyst's surface and it is here that reactions are able to take place. The products then leave the surface, allowing processing of more polluting gases by the catalyst.

Contact process for sulphur trioxide (and subsequently sulphuric acid)

$$2SO_2(g) + O_2(g) \rightleftharpoons 2SO_3(g)$$

- The gaseous equilibrium process is catalysed by solid vanadium(V) oxide, $V_2O_5(s)$.

Catalytic converters (see also page 120)

Many polluting gases, such as carbon monoxide, CO, and nitrogen monoxide, NO, are produced in a car engine.

The catalytic converter has a surface area equivalent to two football pitches! The reacting gases now have a much greater chance of being adsorbed onto the catalyst so that reaction can take place.

These gases are formed by:

- incomplete combustion of hydrocarbons forming CO
- reaction between N_2 and O_2 in a hot car engine forming NO.

These polluting gases are made to react using the catalytic converter in a car's exhaust system. The solid Rh/Pt/Pd catalyst is supported on a ceramic honeycomb which gives a large surface area for catalysis, a feature that greatly improves the effectiveness of many solid heterogeneous catalysts.

The catalytic converter removes much of this pollution in a series of reactions, e.g.

$$2CO(g) + 2NO(g) \longrightarrow 2CO_2(g) + N_2(g)$$

Homogeneous catalysis

A homogeneous catalyst does react but it is then regenerated, to be used again in a chain reaction – the middleman of chemistry.

During homogeneous catalysis, the catalyst does react forming an *intermediate state* with lower activation energy. Subsequently, the catalyst is regenerated, allowing further reaction to take place. Overall the catalyst is not used up, but it is actively involved in the process.

The breakdown of atmospheric ozone, catalysed by Cl· radicals

Chlorine free radicals, Cl· are formed in the stratosphere by the action of ultraviolet radiation on chlorofluorocarbons, CFCs. The gaseous Cl· free radicals catalyse the breakdown of gaseous ozone – the catalysis is homogeneous.

The Cl· free radicals are recycled in a chain reaction. A single Cl· free radical can break down thousands of O_3 molecules.

chain reaction:
$$O_3(g) + Cl·(g) \longrightarrow O_2(g) + ClO·(g)$$
$$O(g) + ClO·(g) \longrightarrow O_2(g) + Cl·(g)$$

overall:
$$O_3(g) + O(g) \longrightarrow 2O_2(g)$$

The O(g) free radicals are formed naturally in the stratosphere by the action of ultraviolet radiation on $O_2(g)$.

See also free-radical substitution of alkanes, p.121.

6.3 Chemical equilibrium

After studying this section you should be able to:

- explain the main features of a dynamic equilibrium
- predict the effects of changes on the position of equilibrium
- know that a catalyst does not affect the position of equilibrium
- understand why a compromise temperature and pressure may be used to obtain an economic yield in industrial processes

LEARNING SUMMARY

Reversible reactions and dynamic equilibrium

AQA	M2	OCR	M3		
EDEXCEL	M2	SALTERS	M2		
NICCEA	M2	WJEC	CH2		
NUFFIELD	M2				

Reversible reactions

Many chemical reactions take place until the reactants are completely used up and such reactions *'go to completion'*.

An example of a reaction that goes to completion is the oxidation of magnesium:

$$2Mg(s) + O_2(g) \longrightarrow 2MgO(s)$$

In a chemical equation,

reactants are on the left-hand side;

products are on the right-hand side.

- The sign '$\longrightarrow$' indicates that the reaction takes place from left to right: this is called the *forward direction*.

Many reactions are **reversible**: they can take place in either direction.

An example of a reaction that is reversible is the formation of ammonia $NH_3(g)$ from nitrogen $N_2(g)$ and hydrogen $H_2(g)$.

$$N_2(g) + 3H_2(g) \rightleftharpoons 2NH_3(g)$$

The sign '$\rightleftharpoons$' indicates that the reaction is reversible.

The reversible reaction:
$N_2(g) + 3H_2(g) \rightleftharpoons 2NH_3(g)$
is used to illustrate dynamic equilibrium throughout this chapter.

This is a homogeneous equilibrium – one in which all reactants and products have the same phase. In this system, all are gases.

Approaching equilibrium

- If nitrogen and hydrogen are mixed together, the reaction can proceed only in the forward direction because no products are yet present:

$$N_2(g) + 3H_2(g) \longrightarrow$$

- As the reaction proceeds, ammonia is formed. The ammonia starts to react in the *reverse direction*, indicated by '$\longleftarrow$', forming nitrogen and hydrogen. At first, there is so little ammonia present that the reverse process takes place extremely slowly:

$$N_2(g) + 3H_2(g) \overset{\longrightarrow}{\longleftarrow} 2NH_3(g)$$

- As more ammonia is formed the reverse process takes place faster. Nitrogen and hydrogen are being used up and the forward reaction slows down. Eventually, the reverse process takes place at the same rate as the forward reaction and **equilibrium** is attained:

$$N_2(g) + 3H_2(g) \rightleftharpoons 2NH_3(g)$$

Dynamic equilibrium

At equilibrium, there is a balance: reactants and products are both present and the reaction *appears* to have stopped.

- Although there is no apparent change, both forward and reverse processes continue to take place – the equilibrium is *dynamic*.

- The forward reaction proceeds at the same rate as the reverse reaction.

In a chemical equation,

reactants are on the left-hand side;

products are on the right-hand side.

A reversible reaction can be approached from either direction and the terms *reactants* and *products* need to be used with caution.

- The concentrations of reactants and products are constant.

The equilibrium position:

- can be reached from either forward or reverse directions
- can only be achieved in a *closed system* – one in which no materials are being added or removed.

Equilibrium may be approached from either direction

Factors affecting equilibrium

AQA	M2	OCR	M3
EDEXCEL	M2	SALTERS	M2
NICCEA	M2	WJEC	CH2
NUFFIELD	M2		

By opening up a closed equilibrium system, conditions can be changed after which the system can be allowed to reach equilibrium again.

The equilibrium position of the system may be altered by the following changes.

- Changing the concentration of a reactant or product.
- Changing the pressure of a gaseous equilibrium.
- Changing the temperature.

The likely effect on the equilibrium position can be predicted using Le Chatelier's Principle.

> *Le Chatelier's Principle* states that if a system in dynamic equilibrium is subjected to a change, processes will occur to minimise this change and to restore equilibrium.
>
> **KEY POINT**

The effect of concentration changes

A change in the concentration of a reactant will alter the rate of the forward direction. A change in the concentration of a product will alter the rate of the reverse direction. Either change results in a shift in the equilibrium position.

$$N_2(g) + 3H_2(g) \rightleftharpoons 2NH_3(g)$$

Change: **Increase** concentration of a reactant: $N_2(g)$ or $H_2(g)$

Change opposed: The equilibrium system will **decrease** the concentration of the reactant by removing it. This is achieved by a shift in equilibrium to the right, forming more $NH_3(g)$.

If the concentration of a reactant is increased, or the concentration of the product is decreased, e.g. by removing some of it, the equilibrium is displaced to the right and more product is obtained.

The effects of changes in concentration for this equilibrium are shown below. These effects will apply to any system in equilibrium.

> increase concentration of reactant OR reduce concentration of product
>
> $$N_2(g) + 3H_2(g) \rightleftharpoons 2NH_3(g)$$
>
> decrease concentration of reactant OR increase concentration of product
>
> **KEY POINT**

The effect of pressure changes

A change in total pressure may alter the equilibrium position of a system involving gases. The direction favoured depends upon the total number of gas molecules on each side of the equilibrium.

$$N_2(g) + 3H_2(g) \rightleftharpoons 2NH_3(g)$$

Change: Increase the total pressure.

Change opposed: The equilibrium system will decrease the pressure by reducing the total number of moles of gas molecules. This is achieved by a shift in equilibrium to the right:

$$N_2(g) \quad + \quad 3H_2(g) \quad \rightleftharpoons \quad 2NH_3(g)$$

| 1 mol | 3 mol | 2 mol |

$$\underbrace{\qquad\qquad\qquad}_{4 \text{ mol}} \qquad\qquad 2 \text{ mol}$$

> **Movement of gas molecules causes pressure. The more gas molecules in a volume, the greater the pressure.**

> **If there are more moles of gaseous reactant than there are moles of gaseous product, an increase in total pressure will displace the reaction to the right.**

KEY POINT

Increasing the pressure also increases the concentration of any gas present and this speeds up the reaction.

The equilibrium position is also changed if:

- at least one of the equilibrium species is a gas
- there are different numbers of gaseous moles of reactants and of products.

The effect of temperature changes

A change in temperature alters the rates of the forward and reverse reactions by different amounts, resulting in a shift in the equilibrium position. The direction favoured depends upon the sign of the enthalpy change.

$$N_2(g) + 3H_2(g) \rightleftharpoons 2NH_3(g) \qquad \Delta H^\ominus = -92 \text{ kJ mol}^{-1}$$

Change: Decrease the temperature.

Change opposed: The equilibrium system will increase the temperature by releasing more heat energy. This is achieved by a shift in equilibrium in the exothermic direction, to the right.

> **An exothermic process is favoured by low temperatures.**
> **An endothermic process is favoured by high temperatures.**

> **An exothermic reaction in one direction is endothermic in the opposite direction.**
> **The value of the enthalpy changes is the same – the sign is different.**

exothermic process favoured by low temperatures

$$\Delta H^\ominus = +92 \text{ kJ mol}^{-1} \xrightarrow{} N_2(g) + 3H_2(g) \rightleftharpoons 2NH_3(g) \qquad \Delta H^\ominus = -92 \text{ kJ mol}^{-1}$$

endothermic process favoured by high temperatures

KEY POINT

Increasing the temperature *speeds* up the reaction and less time is needed to reach equilibrium. However, the equilibrium position may change to produce a lower equilibrium yield of products (see also pages 107–108).

The effect of a catalyst

There is no change in the equilibrium position. However, the catalyst speeds up both forward and reverse reactions and equilibrium is reached quicker.

Equilibria, rates and industrial processes

AQA	M2	OCR	M3
EDEXCEL	M2	WJEC	CH2
NICCEA	M2		

All courses study one or more industrial processes, usually including the Haber and Contact processes.

You should learn outline details of any process on your course.

The principles, however, are common for any similar situation.

For many industrial processes, the overall yield of a chemical product can vary considerably, depending on the conditions used.

In deciding the conditions to use, industrial chemists need to consider:

- the availability of the starting materials required for a process
- the equilibrium conditions required to ensure a good yield
- the rate of reaction which should be fast but manageable
- the cost, which should be as low as possible, taking into account energy, cost of materials and of the chemical plant
- the safety of workers from hazardous chemicals, pressure, temperature, etc.
- any effects on the environment from waste discharges, toxic fumes, etc.

Industrial chemists compare each of these factors to arrive at compromise conditions. Although the equilibrium yield might be optimised by using a very high temperature and pressure, this may prove impracticable for reasons of cost and safety.

The Haber Process for the production of ammonia

AQA	M2	OCR	M3
EDEXCEL	M2	WJEC	CH2
NICCEA	M2		

$$N_2(g) + 3H_2(g) \rightleftharpoons 2NH_3(g) \qquad \Delta H^\ominus = -92 \text{ kJ mol}^{-1}$$

The raw materials

The raw materials used should be readily available. The Haber Process uses:

- the air, as a source of $N_2(g)$
- natural gas, CH_4, and water, H_2O, from which $H_2(g)$ is extracted.

The optimum equilibrium conditions

Consider Le Chatelier's Principle

Optimising the process

These are the ideal conditions to give a maximum equilibrium yield.

- The forward reaction produces fewer moles of gas, favoured by a high pressure.
- The forward reaction producing NH_3 is exothermic, favoured by a low temperature.

The optimum equilibrium conditions for maximum equilibrium yield are:

- **high pressure and low temperature.**

The need for compromise

Compromising

Optimum conditions, reaction rate, feasibility, reality, safety and economics must be considered.

Having arrived at optimum equilibrium conditions, each must be considered to arrive at compromise conditions.

optimum condition	advantages	disadvantages
	equilibrium yield of ammonia is high	energy costs are high – it is expensive to compress gases
HIGH PRESSURE	the concentration of gases is high, increasing the rate	there are considerable safety implications of using very high pressures – vessel walls need to be very thick to withstand high pressures and weaknesses cause danger to workers and potential leakage of chemicals into the environment
LOW TEMPERATURE	equilibrium yield of ammonia is high	the reaction takes place very slowly as few molecules possess the activation energy of the reaction

The use of a catalyst

A catalyst speeds up the rates of both the forward and reverse reactions and hence less time is needed for the reaction to reach equilibrium. The increase in reaction rate allows lower temperatures to be used for a realistic reaction rate. The use of lower temperatures has the added bonus of supporting the optimum conditions for this process, increasing the equilibrium yield of ammonia.

Compromise conditions

Exam papers always test the wider aspects of chemistry to society.

Exam answers need to be sensible and related to relevant issues arising from the process itself. It is pointless to use terms such as 'dangerous', 'toxic', etc. unless substantiated with the reasons why the hazard is present.

The compromise conditions used in the Haber Process are such that:

- the temperature is sufficiently high to allow the reaction to occur at a realistic rate, but not too high to give a minimal equilibrium yield of ammonia
- a high pressure is used but not too high as to be impracticable
- the reaction rate is increased by using an iron catalyst with potassium hydroxide added to promote its activity. This enables the process to take place at lower temperatures, saving on energy costs.

The compromise conditions used (20 000 kPa and 450°C) only yield 15% $NH_3(g)$ which is removed by liquefying at $-40°C$. Unreacted $N_2(g)$ and $H_2(g)$ are recycled.

Economic considerations

Much of the cost of any industrial process goes on energy. Companies carry out research to reduce energy demands. Recent developments mean that the Haber Process can be carried out using:

- lower temperatures (a greater equilibrium yield and less energy for heating)
- lower pressures (less energy required to compress gases).

Progress check

1 What will be the result of an increase in pressure on the following equilibria?
(a) $N_2O_4(g) \rightleftharpoons 2NO_2(g)$
(b) $CO(g) + 2H_2(g) \rightleftharpoons CH_3OH(g)$
(c) $H_2(g) + Br_2(g) \rightleftharpoons 2HBr(g)$

2 What will be the result of an increase in temperature on the following equilibria?
(a) $N_2(g) + O_2(g) \rightleftharpoons 2NO(g)$ $\Delta H^\ominus = +180$ kJ mol^{-1}
(b) $H_2(g) + Br_2(g) \rightleftharpoons 2HBr(g)$ $\Delta H^\ominus = -9.6$ kJ mol^{-1}
(c) $2SO_2(g) + O_2(g) \rightleftharpoons 2SO_3(g)$ $\Delta H^\ominus = -197$ kJ mol^{-1}

3 What are the optimum conditions of temperature **and** pressure for a high yield of the products in the equilibria below?
(a) $CO(g) + 2H_2(g) \rightleftharpoons CH_3OH(g)$ $\Delta H^\ominus = -92$ kJ mol^{-1}
(b) $PCl_5(g) \rightleftharpoons PCl_3(g) + Cl_2(g)$ $\Delta H^\ominus = +124$ kJ mol^{-1}

3 (a) low temperature and high pressure;
(b) high temperature and low pressure.
2 (a) moves to right; (b) moves to left; (c) moves to left.
1 (a) moves to left; (b) moves to right; (c) no change.

6.4 Acids and bases

After studying this section you should be able to:

- *describe and use the Brønsted–Lowry theory of acids and bases*
- *explain the differences between strong and weak acids*
- *describe the typical reactions of an acid in terms of H⁺(aq)*
- *describe the use of ammonia as a base in the preparation of fertilisers*

Acids, bases and alkalis

| NUFFIELD | M1 | SALTERS | M2 |
| OCR | M3 | WJEC | CH2 |

An acid is a proton donor.
A base is a proton acceptor.

The Brønsted–Lowry theory treats acids and bases in terms of proton transfer.

> **KEY POINT**
>
> A Brønsted–Lowry acid is a proton, H⁺, donor
> A Brønsted–Lowry base is a proton, H⁺, acceptor

Acids

A hydrogen ion, H⁺ is simply a hydrogen atom with an electron removed.
Because a hydrogen atom comprises one proton and one electron, removing the electron leaves just a proton.

'H⁺', 'Hydrogen ion' and proton refer to the same thing.

Common acids include:

- strong 'mineral' acids: sulphuric acid, H_2SO_4; hydrochloric acid, HCl; nitric acid, HNO_3
- weak organic acids: ethanoic acid, CH_3COOH; methanoic acid, HCOOH; benzoic acid, C_6H_5COOH; phenol, C_6H_5OH

When added to water, HCl acts as a Brønsted–Lowry acid, donating a proton to water to form the **oxonium** ion, H_3O^+(aq):

$$HCl(aq) + H_2O(l) \longrightarrow H_3O^+(aq) + Cl^-(aq)$$

proton donor — H⁺ → oxonium ion

> **KEY POINT**
>
> All acids generate H⁺(aq) in aqueous solution.

- The oxonium ion, H_3O^+(aq) is the 'active' part of an aqueous acid.
- It is common practice to simply use H⁺(aq) instead of H_3O^+(aq) and the dissociation of HCl can more simply be shown as:

$$HCl(aq) + aq \longrightarrow H^+(aq) + Cl^-(aq)$$

Strong acids

H_2SO_4, HCl and HNO_3 are strong acids because dissociation in water goes virtually to completion.

A **strong** acid such as nitric acid HNO_3 is a **good** proton donor, with near to complete dissociation in water.

$$HNO_3(l) + aq \longrightarrow H^+(aq) + NO_3^-(aq)$$

A **strong** acid **completely dissociates** into ions.

Weak acids

A weak acid only partially dissociates

A **weak** acid such as ethanoic acid CH_3COOH is a **poor** proton donor. The dissociation in water is an equilibrium, with the equilibrium position well to the left-hand side of the equation.

$$CH_3COOH(aq) \rightleftharpoons H^+(aq) + CH_3COO^-(aq)$$

equilibrium position

For every 250 molecules of ethanoic acid, only 1 molecule dissociates and only a small proportion of the potential acidic power is released as H⁺.

A **weak** acid only **partially dissociates** into ions.

Bases

Common bases include:

- metal oxides: magnesium oxide, MgO; copper(II) oxide, CuO
- metal hydroxides: sodium hydroxide, NaOH; iron(III) hydroxide, $Fe(OH)_3$
- ammonia and amines: ammonia, NH_3; methylamine, CH_3NH_2

Alkalis

An alkali, such as NaOH, is a base that dissolves in water forming $OH^-(aq)$ ions:

$$NaOH(s) + aq \longrightarrow Na^+(aq) + OH^-(aq)$$

There are only a few common alkalis.

- strong alkalis: potassium hydroxide, KOH; lithium hydroxide, LiOH
- weak alkalis: ammonia, NH_3; magnesium hydroxide, $Mg(OH)_2$

Typical reactions of an acid

OCR　　M3
NUFFIELD　　M1

Ionic equations can be written for the typical acid reactions. Note the role of $H^+(aq)$ in each of these reactions. By using $H^+(aq)$, the same reaction can be written, irrespective of the acid. (Remember, the 'active part' of an acid is $H^+(aq)$.)

Using hydrochloric acid as an example, typical reactions of an acid are shown below. The ionic equations clearly show the role of $H^+(aq)$.

Acids react with many metals, evolving hydrogen

$$2HCl(aq) + Mg(s) \longrightarrow H_2(g) + MgCl_2(aq)$$
$$2H^+(aq) + Mg(s) \longrightarrow H_2(g) + Mg^{2+}(aq)$$

Acids react with carbonates, evolving carbon dioxide

$$2HCl(aq) + CaCO_3(s) \longrightarrow CaCl_2(aq) + H_2O(l) + CO_2(g)$$
$$2H^+(aq) + CaCO_3(s) \longrightarrow H_2O(l) + CO_2(g) + Ca^{2+}(aq)$$

Acids react with bases (metal oxides), forming water

$$2HCl(aq) + CaO(s) \longrightarrow H_2O(l) + CaCl_2(aq)$$
$$2H^+(aq) + CaO(s) \longrightarrow H_2O(l) + Ca^{2+}(aq)$$

Acids react with alkalis, forming water

$$HCl(aq) + NaOH(aq) \longrightarrow NaCl(aq) + H_2O(l)$$
$$H^+(aq) + OH^-(aq) \longrightarrow H_2O(l)$$

Acid reactions
metal
↓ H^+
H_2
carbonate
↓ H^+
CO_2
base
↓ H^+
H_2O
alkali
↓ H^+
H_2O

Salts

In each of these reactions a salt is formed, in which the active part of the acid, H^+, has been replaced by a metal. Examples of some common salts are shown below.

Salts are always formed in solution from these typical acid reactions. The salt can be isolated by evaporation of water.

acid		salt	example
sulphuric acid	H_2SO_4	sulphate	Na_2SO_4
hydrochloric acid	HCl	chloride	NaCl
nitric acid	HNO_3	nitrate	KNO_3

Ammonia as a base

EDEXCEL　　M2
OCR　　M3

Ammonia NH_3 is a colourless gas, very soluble in water forming a weakly alkaline solution:

$$NH_3(g) + H_2O(l) \rightleftharpoons NH_4^+(aq) + OH^-(aq)$$

Much of the ammonia formed in the Haber Process (see page 107) is neutralised by acids to form ammonium salts, such as ammonium nitrate, shown below.

$$NH_3(aq) + HNO_3(aq) \longrightarrow NH_4NO_3(aq)$$

Ammonium salts such as ammonium nitrate are very soluble in water and have a high nitrogen content. They are extensively used as artificial fertilisers.

Sample question and model answer

1

The curve below represents the distribution of molecular energies at a temperature T_1 for a mixture of gases which react with each other. **A** is the activation energy for the reaction.

Number of molecules with a given energy

T_2

T_1

T_2 is lower than T_1✓
to the left of T_1✓

✓

B✓   A   energy

(a) (i) Label the vertical axis.

(ii) Explain the meaning of the term *activation energy*.

This is the minimum ✓ energy required for a reaction to take place. ✓

(iii) Draw a second curve on the same axes, and label it T_2, for the same mixture at a lower temperature.

(iv) By reference to the curves, state and explain in molecular terms the effect of reducing the temperature on the rate of reaction.

Reduced temperature slows down the reaction rate. ✓
The overall kinetic energy of the molecules is reduced ✓ and fewer molecules now exceed the activation energy. ✓   [9]

(b) The reaction is repeated in the presence of a catalyst.

(i) Mark on the energy axis a possible activation energy B for the catalysed reaction.

(ii) State and explain, in molecular terms, the effect of the catalyst on the rate of reaction.

The catalyst speeds up the reaction rate. ✓ A catalyst allows the reaction to proceed via a different route ✓ with a lower activation energy, B. This means that a larger proportion of molecules now exceeds the activation energy. ✓   [4]

(c) The equation for the industrial manufacture of ammonia is given below.
$$N_2(g) + 3H_2(g) \rightleftharpoons 2NH_3(g) \qquad \Delta H = -92 \text{ kJ mol}^{-1}$$

A temperature of 450°C is used in industry, together with pressures between 200 and about 250 atmospheres. A catalyst is also used.

(i) Explain why the reaction is carried out at a high pressure.
There is a greater equilibrium yield. ✓ The increase in pressure is relieved by reducing the number of gas molecules. ✓ The equilibrium shifts in favour to the right because there are fewer gas molecules to the right. ✓

(ii) Explain why pressures higher than 1000 atmospheres are not used.
High pressures are very costly to generate in terms of energy. ✓

(iii) State **one** advantage and **one** disadvantage of carrying out the reaction at high temperature.
An advantage is that there is a greater reaction rate. ✓
A disadvantage is that the equilibrium yield is reduced. ✓   [6]

[Total: 19]
NEAB Kinetics and Organic Chemistry Q2 June 1995 (modified)

Notice that the distribution curve starts at the origin but it never quite reaches the x-axis, even at high energies. Remember also that the area under the curve is equal to the total number of molecules.

At different temperatures,
the curve changes
the activation energy stays the same.

Note the difference

In the presence of a catalyst,
the activation energy changes
the curve stays the same.

Always pay attention to the words that examiners use.

In parts (i) and (ii), the examiner has used 'Explain'. You need to justify your answer.

In part (iii), the examiner has used 'State'. You only need to 'state' your answer. You waste time here if you 'explain'.

Practice examination questions

1 (a) Define the term *activation energy*. [1]

(b) The endothermic reaction between substances **P** and **Q** can be represented by the following equation.

$$P(g) + Q(g) \longrightarrow R(g) + S(g)$$

(i) Complete the reaction profile for this reaction, shown in **Figure 1**. Indicate and label clearly the activation energy and the enthalpy change for the reaction.

(ii) Also on **Figure 1**, draw and label the reaction profile for the reaction when it is catalysed.

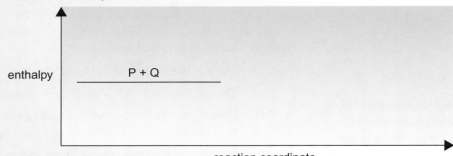

Figure 1

(iii) Explain why a catalyst increases the rate of a chemical reaction. [6]

Figure 2

(c) (i) **Figure 2** shows the distribution of molecular energies for the mixture of reactants in **(b)** at a temperature T_1. Label the axes and, on the same axes, draw the distribution of molecular energies for the same gaseous mixture at a higher temperature T_2.

(ii) In **Figure 2**, the energy corresponding to the activation energy of the reaction (E_A) is shown. Indicate a possible energy corresponding to the activation energy of the catalysed reaction and label it E_A **catalysed**. [5]

[Total: 12] *AEB Paper 1 Q6 Jun 1997 (modified)*

2 A dynamic homogeneous equilibrium is established when hydrogen and carbon dioxide react as shown below:

$$H_2(g) + CO_2(g) \rightleftharpoons H_2O(g) + CO(g)$$

(a) (i) Why is this reaction a *homogeneous equilibrium*?

(ii) State **three** features of a *dynamic equilibrium*. [5]

(b) The equilibrium yield of steam and carbon monoxide increases as the temperature is increased. Determine whether the reaction as written is exothermic or endothermic. Explain your answer. [2]

(c) Explain the effect of an increase in pressure on:

(i) the equilibrium position; (ii) the reaction rate. [4]

[Total 11]

Chapter 7

Organic chemistry

The following topics are covered in this chapter:

- Basic concepts
- Hydrocarbons from oil
- Alkanes
- Alkenes
- Alcohols
- Halogenoalkanes

7.1 Basic concepts

After studying this section you should be able to:

- understand the different types of formula used for organic compounds
- understand the terms saturated and unsaturated hydrocarbon
- understand what is meant by structural isomerism
- recognise the alkanes and alkyl groups
- understand what is meant by a homologous series
- recognise common functional groups
- apply rules for naming simple organic compounds
- understand the difference between homolytic and heterolytic fission
- calculate percentage yields

LEARNING SUMMARY

Types of formula

AQA	M3	OCR	M2
EDEXCEL	M2	SALTERS	M1
NICCEA	M2	WJEC	CH2
NUFFIELD	M1		

In organic chemistry, there are many ways of representing a formula.
For the compound butane, with 4 carbon atoms and 10 hydrogen atoms:

the *empirical* formula is: C_2H_5 — The simplest, whole-number ratio of elements in a compound.

the *molecular* formula is: C_4H_{10} — The *actual* number of atoms of each element in a molecule.

the *structural* formula is: $CH_3CH_2CH_2CH_3$ — The minimal detail for an unambiguous structure.

the *displayed* formula is:

$$H-\underset{\underset{H}{|}}{\overset{\overset{H}{|}}{C}}-\underset{\underset{H}{|}}{\overset{\overset{H}{|}}{C}}-\underset{\underset{H}{|}}{\overset{\overset{H}{|}}{C}}-\underset{\underset{H}{|}}{\overset{\overset{H}{|}}{C}}-H$$

The relative placing of atoms and the bonds between them.

the *skeletal* formula

The carbon skeleton and functional groups only.

Carbon chains

AQA	M3	OCR	M2
EDEXCEL	M2	SALTERS	M1
NICCEA	M2	WJEC	CH2
NUFFIELD	M1		

Hydrocarbons

Hydrocarbons are compounds of carbon and hydrogen only.

- A saturated hydrocarbon has single bonds only.
- An unsaturated hydrocarbon contains a multiple carbon carbon bond.

saturated
single bonds only

unsaturated
contains a double bond

Alkanes

Carbon atoms can bond with other carbon atoms to form an enormous range of compounds with different carbon-chain lengths. The simplest organic compounds are a family of saturated hydrocarbons called the alkanes, shown below.

meth- C
eth- C_2
prop- C_3
but- C_4
pent- C_5
hex- C_6

Number of carbons	Name	Molecular formula	Structural formula
1	methane	CH_4	CH_4
2	ethane	C_2H_6	CH_3CH_3
3	propane	C_3H_8	$CH_3CH_2CH_3$
4	butane	C_4H_{10}	$CH_3CH_2CH_2CH_3$
5	pentane	C_5H_{12}	$CH_3CH_2CH_2CH_2CH_3$
6	hexane	C_6H_{14}	$CH_3CH_2CH_2CH_2CH_2CH_3$

Note the following points.

- The name of the alkane ends with –*ane*.
- The prefixes (*meth-*, *eth-*, …) are used to represent the number of carbon atoms. You will need to use these many times in organic chemistry and they must be learnt.

General formula

Alkanes, C_nH_{2n+2} are saturated.

- The general formula is the simplest algebraic representation for any member in a series of organic compounds.
- The general formula for any alkane is C_nH_{2n+2}, where n = the number of carbon atoms.

Homologous series

Members in a homologous series react similarly.

By studying the reactions of one member of the series, you know how all members in the series are likely to react.

The alkanes are an example of a homologous series with the following features.

- Each successive member differs by $-CH_2-$. You can see this by comparing the formula of each alkane in the table above.
- Each member in a homologous series has the same general formula. The alkanes have the general formula: C_nH_{2n+2}.
- All members in a homologous series have the same functional group and similar chemical reactions.

Note that physical properties, such as boiling point and density, do gradually change as the length of the carbon chain increases (see also page 120).

Isomerism

AQA	M3	OCR	M2
EDEXCEL	M2	SALTERS	M1
NICCEA	M2	WJEC	CH2
NUFFIELD	M1		

In addition to forming long chains, the atoms making up a molecular formula are often arranged differently, forming isomers.

Structural isomerism

Chain isomerism

This is a type of structural isomerism in which the carbon skeleton is different. Butane and methylpropane are chain isomers of C_4H_{10}

Structural isomers are molecules with the same molecular formula but with different structural arrangements of atoms.

The two structural isomers of C_4H_{10} are shown below:

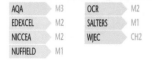

butane   methylpropane

See also *cis–trans* isomerism, p.122.

The number of different structural isomers possible from a molecular formula increases dramatically as the carbon-chain length increases. This is shown in the table below.

molecular formula	number of isomers
C_5H_{12}	3
C_6H_{14}	5
C_7H_{16}	9
C_8H_{18}	18
C_9H_{20}	35
$C_{10}H_{22}$	75
$C_{15}H_{32}$	4,347
$C_{20}H_{42}$	366,319

Look at the different names given to the two isomers.

- The branched isomer, methylpropane, is treated as a side chain attached to a straight-chain alkane.
- Side chains, such as the methyl group CH_3-, are called *alkyl* groups with the general formula of C_nH_{2n+1}.
- An alkyl group can be regarded as an alkane with one hydrogen atom removed (to allow another group to be attached).

The naming of organic compounds is discussed in more detail on page 116.

Notice how the alkyl groups are named and how their formulae are linked to the parent alkane.

You must learn these.

Alkanes and alkyl groups

Number of carbons	Alkane C_nH_{2n+2} Formula	Name	Alkyl group C_nH_{2n+1} Formula	Name
1	CH_4	methane	CH_3-	methyl
2	C_2H_6	ethane	C_2H_5-	ethyl
3	C_3H_8	propane	C_3H_7-	propyl

Functional groups

AQA	M3	OCR	M2
EDEXCEL	M2	SALTERS	M1
NICCEA	M2	WJEC	CH2
NUFFIELD	M1		

Saturated hydrocarbon chains are comparatively unreactive. The reactivity is increased by the presence of a functional group – the reactive part of a carbon compound.

Unreactive carbon chain *alkyl group* C_nH_{2n+1}

Functional group The part responsible for the reactions

Common functional groups

It is essential that you can instantly identify a functional group within a molecule. You can then apply the relevant chemistry.

If you continue to study Chemistry to A Level, you will meet more functional groups.

The main functional groups for AS Chemistry are shown below.

name	functional group	structural formula	general formula	prefix or suffix (for naming)
alkane	C–H	$CH_3CH_2CH_3$ *propane*	C_nH_{2n+2}	-ane
alkene	C=C	$CH_3CH=CH_2$ *propene*	C_nH_{2n}	-ene
halogenoalkane	C–X (X=halogen)	CH_3CH_2Br *bromoethane*	$C_nH_{2n+1}Br$	bromo-
alcohol	C–OH	CH_3CH_2OH *ethanol*	$C_nH_{2n+1}OH$	-ol
aldehyde	—C(=O)H	CH_3CHO *ethanal*	$C_nH_{2n}O$	-al
ketone	R–C(=O)–R	$CH_3COC_2H_5$ *butanone*	$C_nH_{2n}O$	-one
carboxylic acid	—C(=O)OH	CH_3COOH *ethanoic acid*	$C_nH_{2n+1}COOH$	-oic acid

- When studying reactions of a homologous series, the alkyl group is relatively unimportant – it is the functional group that reacts.

- The alkyl group, C_nH_{2n+1} is often represented simply as R–. This shifts the emphasis in the formula towards the reactive part of the molecule, the functional group.

The alcohols, shown below, is a homologous series with the –OH functional group.

- The alcohols have the general formula:
  $C_nH_{2n+1}OH$
- *Any* alcohol can be represented as:
  R–OH

The alcohols
CH_3OH
C_2H_5OH
C_3H_7OH
Same functional group, OH

Naming of organic compounds

AQA	M3	OCR	M2
EDEXCEL	M2	SALTERS	M1
NICCEA	M2	WJEC	CH2
NUFFIELD	M1		

With so many isomers possible, it is important that each has an individual name. The examples below show the steps needed to name an organic compound.

Rule 1

- The name is based upon the longest carbon chain on an alkane.

butane

Rule 2

- Any functional groups and alkyl groups are identified.
- These are then added to the name as a prefix (e.g. chloro-) or suffix (e.g. -ol) (see page 115).

methylbutanol

Rule 3

- If there is more than one possible isomer then the carbon atoms are labelled with numbers. Numbering starts from the end giving the lowest possible numbers for any functional groups and side chains. In this example, numbering from the left would give the incorrect name of 3-methylbutan-4-ol.

2-methylbutan-1-ol

Rule 4

- If there is more than one alkyl or functional group, they are placed in alphabetical order. This rule is not needed for the example above but it does need to be applied to name the compound below.

3-ethyl-2-methylpentane

Types of bond fission

AQA	M3	OCR	M2
EDEXCEL	M2	SALTERS	M2
NICCEA	M2	WJEC	CH2
NUFFIELD	M2		

The breaking of a covalent bond is called bond fission. Reactions of organic compounds involve bond fission followed by the formation of new bonds. Two types of bond fission are possible, homolytic fission and heterolytic fission.

Homolytic fission

These are very important principles, used throughout organic chemistry.

In homolytic fission, bond-breaking produces two species of the same (*homo-*) type:

A:B $\longrightarrow$ A• + •B

free radicals

A free radical is a species with an unpaired electron (see p.121).

- In the example above, a covalent bond breaks so that one of the bonding electrons goes to each of **A** and **B**.
- Homolytic fission forms two free-radicals.

Heterolytic fission

Homolytic fission
⟶ free radicals
Heterolytic fission
⟶ ions

In heterolytic fission, bond-breaking produces two species of different (*hetero-*) type:

$$\textbf{A:B} \longrightarrow \textbf{A:}^- + \textbf{B}^+ \qquad or \qquad \textbf{A:B} \longrightarrow \textbf{A}^+ + \textbf{:B}^-$$
$$\quad ions \qquad\qquad\qquad\qquad\qquad\qquad ions$$

- In the example above, a covalent bond breaks so that both the bonding electrons go to either **A** or **B**.
- Heterolytic fission forms oppositely-charged ions.

Percentage yields

| EDEXCEL | M2 | NUFFIELD | M1 |
| NICCEA | M2 | OCR | M2 |

Organic reactions typically produce a lower yield than expected from the balanced equation. The yield is usually expressed as a percentage yield:

> **KEY POINT**
>
> $$percentage\ yield = \frac{actual\ yield}{theoretical\ yield} \times 100\%$$

Example

5.2 g of 1-bromobutane is formed by reacting 5.0 g of butan-1-ol with sodium bromide and concentrated sulphuric acid. Find the percentage yield of 1-bromobutane.

equation	$C_4H_9OH + NaBr + H_2SO_4$	⟶	$C_4H_9Br + H_2O + NaHSO_4$
moles	1 mol	⟶	1 mol
reacting masses	74 g	⟶	137 g
	1 g	⟶	$\frac{137}{74}$ g
	5.0 g	⟶	$5 \times \frac{137}{74}$ g

Factors that can contribute to a low yield:

Organic reactions often do not go to completion.

An organic compound often reacts to produce a mixture of products.

The purification stages result in loss of some of the desired product.

∴ 5.0 g of C_4H_9OH produces a theoretical yield of 9.3 g C_4H_9Br.

Mass of C_4H_9Br that forms = 5.2 g.

$$percentage\ yield = \frac{actual\ yield}{theoretical\ yield} \times 100\% = \frac{5.2}{9.3} \times 100 = 56\%$$

Progress check

C_6H_{14} has 5 structural isomers. Show the structural and displayed formula for each of these isomers and name them.

7.2 Hydrocarbons from oil

After studying this section you should be able to:

- *understand how hydrocarbon fractions are separated from crude oil*
- *describe how low demand fractions are processed into hydrocarbons of higher demand*

LEARNING SUMMARY

Useful products from crude oil

AQA	M3	OCR	M2
NICCEA	M2	SALTERS	M1
NUFFIELD	M2	WJEC	CH2

Crude oil is a fossil fuel, formed from the decay of sea creatures over millions of years. It is a complex mixture of hydrocarbons, containing mainly alkanes. Crude oil cannot be used directly but it is the source of many useful products. The first stages in the processing of crude oil are described below.

Fractional distillation

Crude oil is heated and passed into a fractionating tower, separating the complex mixture into *fractions*. Note that the fractions are not pure and contain mixtures of hydrocarbons within a range of boiling points. The fractionating tower, shown below, shows the temperature gradient which separates the fractions according to their boiling points (see also page 120).

Fractions with the lowest boiling points are collected at the top of the tower.

On descending the tower, the temperature rises and the boiling points increase.

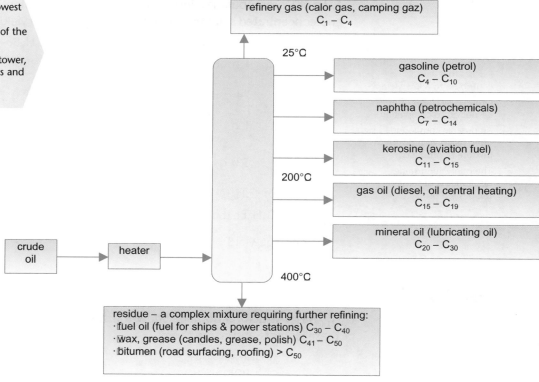

refinery gas (calor gas, camping gaz) $C_1 - C_4$

25°C

gasoline (petrol) $C_4 - C_{10}$

naphtha (petrochemicals) $C_7 - C_{14}$

kerosine (aviation fuel) $C_{11} - C_{15}$

200°C

gas oil (diesel, oil central heating) $C_{15} - C_{19}$

mineral oil (lubricating oil) $C_{20} - C_{30}$

crude oil → heater →

400°C

residue – a complex mixture requiring further refining:
- fuel oil (fuel for ships & power stations) $C_{30} - C_{40}$
- wax, grease (candles, grease, polish) $C_{41} - C_{50}$
- bitumen (road surfacing, roofing) > C_{50}

Cracking

Cracking is the starting point for the manufacture of many organics.

Cracking is the breaking down of an unsaturated hydrocarbon into smaller hydrocarbons.

The purpose of cracking is to produce high demand hydrocarbons:

- short-chain alkanes for use in petrol
- alkenes, as a feedstock for a wide range of organic chemicals, including polymers (see also page 125).

Thermal cracking heats the *naphtha fraction* with steam at a high temperature (about 800°C) and high pressure. This forms a mixture of straight-chain alkanes and alkenes (mainly ethene) with a small proportion of branched and cyclic hydrocarbons. Some hydrogen is also produced.

The following equations summarise the overall process of cracking.

| Thermal cracking breaks C-C bonds by homolytic fission forming free radicals.

Catalytic cracking breaks C-C bonds by heterolytic fission forming ions. |

$$C_{10}H_{22} \longrightarrow C_8H_{18} + C_2H_4$$
$$C_{10}H_{22} \longrightarrow C_6H_{14} + 2C_2H_4$$
$$\text{alkane} \longrightarrow \text{alkane} + \text{alkene}$$

There are many more equations possible but all must produce both an alkane and an alkene.

Catalytic cracking processes heavy long-chain fractions obtained in larger quantities than required. A mixture of the vapourised fraction and a zeolite catalyst are reacted at about 450°C using a slight pressure only. The process forms a higher proportion of branched and cyclic hydrocarbons than thermal cracking (see also reforming and isomerisation below). These are used for petrol.

Reforming

The use of leaded petrol has been phased out in the UK.

Unleaded petrol and lead replacement petrol require a higher proportion of branched and cyclic hydrocarbons for effective combustion. This has increased demand for reformed alkanes.

Reforming uses heat and pressure to convert unbranched fractions into cycloalkanes (e.g. cyclohexane) and arenes (e.g. benzene). These products are used in petrol and as a feedstock for a wide range of organic chemicals including many pharmaceuticals and dyes.

Examples

cyclohexane

benzene

Isomerisation

Isomerisation converts unbranched hydrocarbons into branched hydrocarbons, needed for petrol.

Example

Progress check

1 Write down the formula of an alkane present in each of the main fractions obtained from crude oil.
2 Construct equations for two possible reactions taking place during the cracking of $C_{14}H_{30}$.

7.3 Alkanes

After studying this section you should be able to:

- *understand why different alkanes have different melting and boiling*
- *understand that combustion of alkanes provides useful energy*
- *describe the substitution reactions of alkanes with halogens*

Alkanes

General formula: C_nH_{2n+2}

AQA	M3	OCR	M2
EDEXCEL	M2	SALTERS	M1
NICCEA	M2	WJEC	CH2
NUFFIELD	M2		

Alkanes are generally unreactive. Alkanes contain only C–H and C–C bonds, which are relatively strong and difficult to break. The similar electronegativities of carbon and hydrogen give molecules which are non-polar. Alkanes are the typical 'oils' used in many non-polar solvents and they do not mix with water.

Melting points and boiling points of alkanes

Boiling points increase with increasing chain length because of the increasing strength of the intermolecular forces. This allows the alkane mixture in crude oil to be separated by fractional distillation.

> The same principle can be applied to any homologous series.

Increasing chain length

m.pt and b.pt increase

more surface of contact between molecules

greater van der Waals' forces between molecules

C_3H_8 → C_5H_{12}

For hydrocarbons of similar molecular mass, branched alkanes have lower boiling points because there are fewer interactions possible between molecules:

> The surface contact between molecules is important because it allows interaction between electrons. Remember that van der Waals' forces result from fluctuations in electron density (see p.55).

Isomers of C_4H_{10}

m.pt and b.pt increase

more surface of contact between molecules

greater van der Waals' forces between molecules

Combustion of alkanes

AQA	M3	OCR	M2
EDEXCEL	M2	SALTERS	M1
NICCEA	M2	WJEC	CH2
NUFFIELD	M2		

Combustion with oxygen is the most important reaction of alkanes giving their immediate use as fuels:

natural gas: $\quad CH_4 + 2O_2 \longrightarrow CO_2 + 2H_2O$

petrol: $\quad C_8H_{18} + 12\frac{1}{2}O_2 \longrightarrow 8CO_2 + 9H_2O$

> Although generally unreactive, alkanes do react with oxygen and the halogens.

The burning of fossil fuels such as alkanes provides society with many environmental problems, some of which are listed below.

- Petroleum fractions contain sulphur impurities. Unless removed before combustion, sulphur oxides are formed from which acid rain (H_2SO_3 and H_2SO_4) is produced.

> Remember that reactions with oxygen produce oxides.

- Toxic gases such as carbon monoxide, nitrogen oxides and unburnt hydrocarbons are present in car emissions. The development of catalytic converters has helped to remove these polluting gases (see also page 103).

> These environmental issues are often tested in exams.

- Excessive combustion of fossil fuels increases carbon dioxide emissions contributing to the 'Greenhouse Effect'.

Substitution of alkanes with halogens

AQA	M3	OCR	M2
EDEXCEL	M2	SALTERS	M2
NICCEA	M2	WJEC	CH2
NUFFIELD	M2		

> A *substitution reaction* involves the swapping over of one species for another.

Alkanes are substituted by halogens, such as chlorine and bromine.

$$CH_4 + Cl_2 \longrightarrow CH_3Cl + HCl$$

This reaction only takes place in the presence of ultraviolet radiation, which produces highly reactive free radicals.

> A free radical, such as $Cl\bullet$, $CH_3\bullet$,
> • is a highly reactive species with an unpaired electron
> • reacts by pairing of an unpaired electron
> • is often involved in chain reactions.

Mechanism of free-radical substitution (not EDEXCEL)

Initiation

> Make sure that you learn this mechanism – these are easy marks in exams if you do!

In this stage, the reaction starts.

Ultraviolet radiation provides energy to break Cl–Cl bonds homolytically producing chlorine free-radicals, $Cl\bullet$

$$Cl_2 \longrightarrow 2Cl\bullet$$

Propagation

In this stage, the reaction products are made.

Free radicals are recycled in a chain reaction

> The chlorine free-radicals catalyse the reaction.
> See also the breakdown of ozone, p.103.

$$CH_4 + Cl\bullet \longrightarrow CH_3\bullet + HCl$$
$$CH_3\bullet + Cl_2 \longrightarrow CH_3Cl + Cl\bullet$$

Termination

In this stage, free radicals react together and are removed from the reaction mixture.

$$Cl\bullet + Cl\bullet \longrightarrow Cl_2$$
$$CH_3\bullet + Cl\bullet \longrightarrow CH_3Cl$$
$$CH_3\bullet + CH_3\bullet \longrightarrow CH_3CH_3$$

Impure products

> This is answered poorly in exams.

Further substitution of the reaction products is possible, producing a mixture of products:

$$CH_3Cl \xrightarrow{Cl\bullet} CH_2Cl_2 \xrightarrow{Cl\bullet} CHCl_3 \xrightarrow{Cl\bullet} CCl_4$$

Progress check

1. Write an equation for the complete combustion of butane.
2. Butane reacts with chlorine in a substitution reaction.
   (a) What conditions are essential for this reaction? Explain your answer.
   (b) Write an equation for this reaction.
   (c) Write the structural formula of 2 possible monosubstituted products.
   (d) What is the formula for the final product obtained from complete substitution?

1 $C_4H_{10}(g) + 6\tfrac{1}{2}O_2(g) \longrightarrow 4CO_2(g) + 5H_2O(l)$
2 (a) uv; to generate free radicals from Cl_2
   (b) $C_4H_{10} + Cl_2 \longrightarrow C_4H_9Cl + HCl$
   (c) Any 2 isomers of C_4H_9Cl, e.g. $CH_3CH_2CH_2CH_2Cl$ and $CH_3CH_2CHClCH_3$
   (d) C_4Cl_{10}

7.4 Alkenes

After studying this section you should be able to:

- recognise the nature of cis-trans isomerism in alkenes
- understand that the double bond in alkenes takes part in addition reactions with electrophiles
- understand the use of alkenes in the production of organic compounds
- understand what is meant by addition polymerisation

Alkenes

General formula: C_nH_{2n}

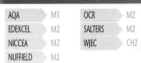

AQA	M3	OCR	M2
EDEXCEL	M2	SALTERS	M2
NICCEA	M2	WJEC	CH2
NUFFIELD	M2		

Alkenes are unsaturated compounds with a C=C double bond. The high electron density of the double bond makes alkenes more reactive than alkanes.

Naming of alkenes

Positional isomerism

A type of structural isomerism in which the functional group is in a different position on the carbon skeleton.

The position of the double bond is indicated using one number only, as shown in the example below for two isomers of C_4H_8.

but-1-ene   but-2-ene

Only one number is needed. in but-1-ene, a double bond starting at carbon-1 must finish at carbon-2

What is a double bond?

The coverage here is greatly simplified but sufficient for AS Chemistry.

In saturated compounds, the C–H and C–C single bonds are σ-bonds. A σ-bond is on the C–H or C–C axis, formed by overlap of s- and p- atomic orbitals.

In unsaturated compounds, the C=C double bond comprises a σ- and a π-bond.

σ- and π-bonds are *molecular orbitals* – these bond together the atoms in a molecule.

A π-bond is above and below the C–C axis, formed by overlap of atomic p-orbitals. The π-bond introduces an area of high electron density in the molecule, shown in the diagram of ethene below.

overlap forms a

p-orbitals

σ-bond

π-bond

The C=C bond
planar molecule
3 σ-electron pairs around each carbon atom.
Bond-angle is approximately 120°

cis-trans (or geometric) isomerism

See also structural isomerism, p.114.

Single C–C bonds can rotate. The C=C double bond, however, restricts rotation and prevents groups from moving from one side of the double bond to the other. This gives rise to *cis-trans* (or geometric) isomerism.

No rotation possible about a C=C bond.

The *cis-trans* isomers of the but-2-ene are shown here:

cis-isomer
(groups on same side)

trans-isomer
(groups on opposite sides)

These are relatively easy exam marks. Make sure that you learn this thoroughly.

Note that, for *cis-trans* isomerism, there must be:

- a C=C double bond
- two **different** groups attached to **each** carbon end of the double bond.

Addition reactions of alkenes

AQA	M3	OCR	M2
EDEXCEL	M2	SALTERS	M2
NICCEA	M2	WJEC	CH2
NUFFIELD	M2		

An addition reaction of an alkene involves the opening of the double bond with the formation of a saturated addition product.

> **KEY POINT**
>
> An *addition reaction* involves two species adding together to make one.
>
> Addition usually converts an unsaturated compound into a saturated compound.

This is made possible by the high electron density of the π-bond which attracts electrophiles.

> **KEY POINT**
>
> An *electrophile*, such as Br_2, HBr, H_2SO_4, NO_2^+:
> * is an electron-deficient species
> * 'attacks' an electron-rich carbon atom or double bond by accepting a pair of electrons.

This type of reaction is called electrophilic addition.

Addition of bromine

The addition reaction between an alkene and bromine, in which the orange colour of bromine is decolourised, is used as a test for unsaturation.

Mechanism of electrophilic addition (not EDEXCEL)

Be careful with the direction of the curly arrows.

Don't get confused by the charges and partial charges within this mechanism.

double bond of alkene induces a dipole on Br_2

electrophilic attack

Remember that a curly arrow shows the movement of an electron pair.

> **KEY POINT**
>
> In mechanisms, *a curly arrow* is used to show the movement of a pair of electrons.
> The movement of the electron pair involves either:
> * the formation of a new covalent bond or
> * the heterolytic fission of a covalent bond (see page 117).
> The *curly arrow* always goes from an electron pair (or double bond).

Addition using acidified manganate(VII)

Note that [O] is used in equations to represent an oxidising agent. Many candidates in exams get this reaction wrong by incorrectly adding Mn. $KMnO_4$ is used as a source of manganate (VII) ions.

Another useful test for unsaturation is the use of cold, dilute acidified manganate(VII). This oxidises the double bond of the alkene with the formation of a diol. The purple colour of manganate(VII) is decolourised.

ethane-1,2-diol

Reduction with hydrogen

> Reduction of a organic compound involves loss of oxygen OR gain of hydrogen (accompanied by a gain of electrons).

Alkenes are reduced when treated with hydrogen gas in the presence of a nickel catalyst at 150°C. The process of adding hydrogen across a double bond is sometimes referred to as hydrogenation.

Addition of hydrogen across C=C double bond

Further addition reactions of ethene

Hydrogenation of C=C is used in the production of solid margarine from unsaturated liquid vegetable oils. Saturated fats are solid and the amount of hydrogenation can be controlled to give the correct texture of margarine.

For more details:

polymerisation, see p.125;

preparation of alcohols, see also p.128 ;

Addition reactions of alkenes are useful in organic synthesis. Some further addition reactions of ethene are shown in the flowchart below.

Although ethanol is made industrially using steam and phosphoric acid H_3PO_4, an alternative synthesis uses:

- addition with concentrated sulphuric acid at room temperature
  $$C_2H_4 + H_2SO_4 \longrightarrow CH_3CH_2OSO_3H$$
- followed by hydrolysis
  $$CH_3CH_2OSO_3H + H_2O \longrightarrow CH_3CH_2OH + H_2SO_4$$

Addition reactions of unsymmetrical alkenes

In the addition of HBr to propene $CH_3CH=CH_2$, two products are possible:

- the secondary bromoalkane $CH_3CHBrCH_3$ as the major product
- the primary bromoalkane $CH_3CH_2CH_2Br$ as the minor product.

Mechanism (AQA and WJEC only)

The driving force for the formation of $CH_3CHBrCH_3$ is the intermediate carbonium ion, shown in the mechanism below.

primary carbonium ion

electrophilic attack

secondary carbonium ion

The reaction with H_2SO_4 proceeds via a similar mechanism. H_2SO_4 can be treated as $H–OSO_2OH$.

The secondary carbonium ion is more stable with 2 alkyl groups feeding electrons towards the positive charge.
The alternative primary carbonium ion has only 1 alkyl group to stabilise the charge.

minor product

major product

Addition polymerisation of alkenes

AQA	M3	OCR	M2
EDEXCEL	M2	SALTERS	M2
NICCEA	M2	WJEC	CH2
NUFFIELD	M2		

An addition polymer is a long-chain molecule with a high molecular mass, made by joining together many small molecules called monomers:

MANY MONOMERS $\longrightarrow$ SINGLE POLYMER

n molecules $\longrightarrow$ 1 single molecule

- The monomer used is an unsaturated alkene containing a double C=C bond.
- The addition polymer formed is a saturated compound without a double bond.

Many different addition polymers can be formed by using different alkene monomer units, based on the ethene molecule.

The equations below shows the formation of poly(ethene), poly(chloroethene) (*pvc*) and poly(phenylethene) (*polystyrene*).

> You should be able to draw a short section of a polymer given the monomer units (and *vice versa*).

- Notice that the principle is the same for each addition polymerisation.

ethene → *poly(ethene)*

> Make sure that you can show the repeat unit of a polymer.

- Each of the polymer structures shows the *repeat unit* in brackets. This is repeated thousands of times in each polymer molecule.

chloroethene → *poly(chloroethene)*

- It is important to show the repeat unit with 'side-links' to indicate that both sides are attached also to other repeat units.

phenylethene → *poly(phenylethene)*

Properties of monomers and polymers

- The monomers are volatile liquids or gases.
- Polymers are solids.
- This difference in physical properties is explained by increased van der Waals' forces between the much larger polymer molecules (see page 55).

Problems with disposal

- Addition polymers are non-biodegradable and take many years to break down.
- Disposal by burning can produce toxic fumes (e.g. depolymerisation produces monomers; dioxins from combustion of chlorinated polymers).

> Rather than simply disposing of waste polymers, there is now movement towards recycling, combustion for energy production, and use as a feedstock for cracking.
>
> This eliminates problems with disposal and helps to preserve finite energy resources.

Progress check

1 Write the structural formula for the organic product for the reaction of but-2-ene with the following:
   (a) Br_2;   (b) H_2/Ni;   (c) HBr;   (d) H_2O/H_3PO_4.

2 Show the repeat unit for the addition polymer formed from propene.

2 (repeat unit structure shown)

1 (a) $CH_3CHBrCHBrCH_3$ (b) $CH_3CH_2CH_2CH_3$ (c) $CH_3CH_2CHBrCH_3$
   (d) $CH_3CH_2CHOHCH_3$.

Epoxyethane, $(CH_2)_2O$

$$H_2C \overset{O}{\overset{\triangle}{-}} CH_2$$

Epoxyethane is an important intermediate used by industry during the production of antifreeze, polyesters, solvents, plasticisers and detergents.

Production of epoxyethane

Epoxyethane is produced commercially by direct synthesis of ethene, with either oxygen or air, in the presence of a silver-based catalyst.

> Epoxyethane is prepared by partial oxidation of ethene.

$$\underset{\substack{H \\ }}{\overset{\substack{H \\ }}{C}} = \underset{\substack{H \\ }}{\overset{\substack{H \\ }}{C}} \quad + \quad \tfrac{1}{2}O_2 \quad \xrightarrow[\text{1–2 MPa}]{\text{Ag / } O_2 \text{ 250–300°C}} \quad H_2C \overset{O}{\overset{\triangle}{-}} CH_2 \quad + \quad H_2O$$

The catalyst is present as a finely-divided layer on a support such as alumina. Epoxyethane is purified by fractional distillation.

Reactions of epoxyethane

> **Hazards from preparing and handling epoxyethane**
>
> Epoxyethane is a colourless gas, b.p. 10 °C, which is both flammable and explosive.
>
> Epoxyethane is also toxic and may cause irritation to the respiratory system and neurological effects.

The three-membered ring in epoxyethane is strained making the molecule highly reactive towards nucleophiles (see page 130) such as water and alcohols. Reaction brings about ring opening.

Hydrolysis of epoxyethane

Much of the epoxyethane produced is converted into ethane-1,2-diol which is used directly as antifreeze and indirectly to make polyesters such as *Terylene*. In industry, this is done by treating epoxyethane with an excess of water at 60°C in the presence of sulphuric acid.

$$H_2C \overset{O}{\overset{\triangle}{-}} CH_2 \quad + \quad H_2O \quad \xrightarrow[\text{H}_2\text{SO}_4 \text{ catalyst}]{\text{excess H}_2\text{O / 60°C}} \quad \underset{\substack{| \\ CH_2 - CH_2}}{\overset{OH \quad OH}{|}}$$

The resulting aqueous solution is concentrated to 70% by evaporation and then purified by fractional distillation.

The properties of ethane-1,2-diol make it ideal for antifreeze:

- low melting point (–12°C)
- high boiling point (198°C)
- complete miscibility with water (with which it readily forms hydrogen bonds).

Ethane-1,2-diol is also used in the manufacture of polyesters, such as *Terylene*.

7.5 Alcohols

After studying this section you should be able to:

- *understand the polarity and physical properties of alcohols*
- *describe how ethanol is prepared*
- *describe oxidation reactions of alcohols*
- *describe nucleophilic substitution reactions of alcohols*
- *describe esterification and dehydration of alcohols*
- *describe tests for the presence of the hydroxyl group*

Alcohols

General formula: $C_nH_{2n+1}OH$

AQA	M3	OCR	M2
EDEXCEL	M2	SALTERS	M2
NICCEA	M2	WJEC	CH2
NUFFIELD	M1		

Types and naming of alcohols

The functional group in alcohols is the **hydroxyl** group, C–OH.

Alcohols can be classified as primary, secondary or tertiary, depending on how many alkyl groups are bonded to **C**–OH. You can see how these types of alcohols are named in the diagrams below.

primary alcohol
1 alkyl group attached to C–OH

secondary alcohol
2 alkyl groups attached to C–OH

tertiary alcohol
3 alkyl groups attached to C–OH

propan-1-ol

propan-2-ol

methylpropan-2-ol

Polarity of alcohols

> The properties of alcohols are dominated by the hydroxyl group, C–OH.

Carbon, oxygen and hydrogen have different electronegativities, and alcohols have polar molecules:

Oxygen is more electronegative than both carbon and hydrogen – both carbon and hydrogen are electron deficient.

The polarity produces electron-deficient carbon and hydrogen atoms, indicated in the diagram above. Depending upon the reagents used, alcohols can react by breaking either the C–O or O–H bond.

Physical properties of alcohols

The –OH group dominates the physical properties of short-chain alcohols.

Hydrogen bonding takes place between alcohol molecules, resulting in:

- higher melting and boiling points than alkanes of comparable M_r
- solubility in water.

The solubility of alcohols in water decreases with increasing carbon chain length as the non-polar contribution to the molecule becomes more important.

hydrogen bonding between ethanol molecules

miscible with water
large polar contribution

insoluble in water
most of the molecule is a non-polar carbon chain

polar head

non-polar tail

polar head

non-polar tail

Preparation of ethanol

AQA　M3　NICCEA　M2
EDEXCEL　M2　OCR　M2

Ethanol finds widespread uses: in alcoholic drinks, as a solvent in the form of methylated spirits, and as a fuel. There are two main methods for its production.

Fermentation of sugars (for alcoholic drinks)

Yeast is added to an aqueous solution containing sugars.

E.g. $$C_6H_{12}O_6(aq) \longrightarrow 2C_2H_5OH(aq) + 2CO_2(g)$$

Hydration of ethene (for industrial alcohol)

Ethene and steam are passed over a phosphoric acid catalyst at 330°C under high pressure (see page 124).

$$C_2H_4(g) + H_2O(g) \longrightarrow C_2H_5OH(l)$$

The preparation method used for ethanol may depend on the raw materials available.

An oil-rich country may use oil to produce ethene, then ethanol. However, this method uses up finite oil reserves.

In a country without oil reserves (especially in hot climates), sugar may provide a renewable source for ethanol production.

Combustion of alcohols

AQA　M3　NUFFIELD　M1
EDEXCEL　M2　OCR　M2
NICCEA　M2　SALTERS　M2

Alcohols such as ethanol and methanol are used as fuels, making use of combustion.

$$C_2H_5OH + 3O_2 \longrightarrow 2CO_2 + 3H_2O$$

Ethanol is used as a petrol substitute in countries with limited oil reserves.

Methanol is used as a petrol additive in the UK to improve combustion of petrol. It also has increasing importance as a feedstock in the production of organic chemicals.

Oxidation of alcohols

AQA　M3　OCR　M2
EDEXCEL　M2　SALTERS　M2
NICCEA　M2　WJEC　CH2
NUFFIELD　M1

The oxidation of different alcohols is an important reaction in organic chemistry. It links alcohols with aldehydes, ketones and carboxylic acids, shown below.

aldehyde, RCHO　　　ketone, RCOR'　　　carboxylic acid, RCOOH

- Aldehydes and ketones both contain the carbonyl group C=O.
- Carboxylic acids contain the carboxyl group COOH.

The stages of oxidation are shown below:

hydroxyl group C–OH $\longrightarrow$ carbonyl group C=O $\longrightarrow$ carboxyl group COOH

Oxidation of primary alcohols

A primary alcohol can be oxidised to an aldehyde and then to a carboxylic acid.

> Oxidation of an organic compound involves gain of oxygen OR loss of hydrogen (accompanied by loss of electrons from the organic compound).

KEY POINT

The stages of oxidation are shown below:

primary alcohol   aldehyde   carboxylic acid

This is carried out using an oxidising agent. The oxidising agent used is a mixture of concentrated sulphuric acid, H_2SO_4 (source of H^+) and potassium dichromate, $K_2Cr_2O_7$ (source of $Cr_2O_7^{2-}$).

For balanced equations, the oxidising agent can be shown simply as [O].

- By heating and distilling the product immediately, oxidation can be stopped at the aldehyde stage.

$$CH_3CH_2OH + [O] \longrightarrow CH_3CHO + H_2O$$

- By refluxing with an excess of the oxidising agent, further oxidation takes place to form the carboxylic acid.

$$CH_3CH_2OH + 2[O] \longrightarrow CH_3COOH + H_2O$$

Oxidation of secondary alcohols

The orange dichromate ions, $Cr_2O_7^{2-}$, are reduced to green Cr^{3+} ions.

By heating with $H^+/Cr_2O_7^{2-}$ a secondary alcohol can be oxidised to form a ketone. No further oxidation normally takes place.

secondary alcohol $\longrightarrow$ ketone

secondary alcohol   ketone

The equation for the oxidation of propan-2-ol is shown below.

$$CH_3(CHOH)CH_3 + [O] \longrightarrow CH_3COCH_3 + H_2O$$

Oxidation of tertiary alcohols

Tertiary alcohols are not oxidised under normal conditions.

tertiary alcohol

Distinguishing between aldehydes and ketones (AQA and Nuff only)

The presence of an aldehyde can easily be detected using two tests.

Tollens' reagent is a solution of silver nitrate (as a source of Ag^+ ions) in ammonia.
$Ag^+ \longrightarrow Ag$

Fehling's (or Benedict's) solution is an alkaline solution of Cu^{2+} ions.
$Cu^{2+} \longrightarrow Cu^+$

Heat aldehyde with Tollens' reagent
- Tollens' reagent $\longrightarrow$ silver mirror

Heat aldehyde with Fehling's (or Benedict's) solution
- Benedict's / Fehling's solution $\longrightarrow$ brick-red precipitate of Cu_2O

Ketones cannot be oxidised and **do not** react with the above solutions and this provides an easy method for distinguishing between an aldehyde and a ketone.

Reduction of carbonyl compounds

WJEC CH2

This is the reverse of the oxidation of alcohols shown above. Alcohols can be prepared by reduction of carbonyl compounds using a suitable reducing agent such as sodium borohydride, $NaBH_4$, in dry ether (at room temperature).

For balanced equations, the reducing agent can be shown simply as [H].

CH_3CHO	+	$2[H]$	$\longrightarrow$	CH_3CH_2OH
aldehyde				primary alcohol
$C_2H_5COCH_3$	+	$2[H]$	$\longrightarrow$	$C_2H_5CH(OH)CH_3$
ketone				secondary alcohol

Substitution reactions of alcohols

EDEXCEL M2 OCR M2

NICCEA M2 SALTERS M2

The electron-deficient carbon atom of the polar $C^{\delta+}-O^{\delta-}$ bond of alcohols attracts *nucleophiles*, allowing a substitution reaction to take place in which the nucleophile replaces the OH group. This is called nucleophilic substitution (see also halogenoalkanes, page 132).

> A nucleophile, such as OH^-, Br^-, H_2O, NH_3, CH_3OH, CN^-
> - is an electron-rich species
> - 'attacks' an electron-deficient carbon atom by donating a pair of electrons.

KEY POINT

An alternative method is to use phosphorus halides.

For chloroalkanes, PCl_5 is used (see p.131).

For iodoalkanes, phosphorus and iodine is used.

Reaction with bromide ions, Br⁻ in acid, H⁺

$$C_2H_5OH + Br^- + H^+ \longrightarrow C_2H_5Br + H_2O$$

The polarity of the C–OH bond allows a reaction with nucleophiles such as Br^-, which attack the electron-deficient carbon atom. The presence of an acid, H^+, is essential as it initially reacts with the alcohol forming a better leaving group:

Use an acid with a nucleophile in substitution reactions of alcohols.

'H_2O' is a better 'leaving group' than OH

- Most reactions of alcohols require the presence of an acid.

Dehydration of alcohols

AQA M3 OCR M2

EDEXCEL M2 SALTERS M2

NUFFIELD M1

When an alcohol is refluxed with a concentrated acid catalyst such as sulphuric acid H_2SO_4, or phosphoric acid H_3PO_4, an elimination reaction takes place: water is eliminated with formation of an alkene.

$$C_2H_5OH \longrightarrow C_2H_4 + H_2O$$

Use hot concentrated acid for elimination of water from an alcohol.

> An *elimination* reaction involves one species breaking up into two – the opposite to addition.
> Elimination usually converts a saturated compound into an unsaturated compound.

KEY POINT

Elimination (1 ⟶ 2) is the opposite process to addition (2 ⟶ 1).

Formation of esters

NICCEA   M2   SALTERS   M2
OCR   M2

Esters are formed by reaction of a carboxylic acid with an alcohol.

Esters are formed by reaction of an alcohol with a carboxylic acid in the presence of an acid catalyst (e.g. concentrated sulphuric acid).

$$\text{carboxylic acid} + \text{alcohol} \longrightarrow \text{ester} + \text{water}$$

The esterification of ethanol by ethanoic acid is shown below:

$$CH_3COOH + CH_3OH \longrightarrow CH_3COOCH_3 + H_2O$$

Tests for alcohols

EDEXCEL   M2   NUFFIELD   M1
NICCEA   M2   OCR   M2

Sodium releases H_2 from any compound with a hydroxyl group: alcohols, carboxylic acids (and water).

Reaction with sodium

Alcohols react with sodium, releasing hydrogen gas. The reaction involves breaking of the O–H bond of the alcohol.

$$2C_2H_5OH + 2Na \longrightarrow 2C_2H_5O^-Na^+ + H_2$$
$$\text{sodium ethoxide}$$

Reaction with phosphorus pentachloride PCl$_5$

(EDEXCEL and NICCEA only)

Alcohols react with PCl$_5$, releasing hydrogen chloride gas which forms misty fumes in air.

$$C_2H_5OH + PCl_5 \longrightarrow C_2H_5Cl + HCl + POCl_3$$

In the presence of ammonia, HCl fumes form dense white fumes of ammonium chloride:

$$NH_3 + HCl \longrightarrow NH_4Cl$$

Progress check

1 Why does ethanol dissolve in water?

2 Write equations to show possible oxidations for the following:
(a) butan-1-ol;   (b) butan-2-ol.

3 What are the 2 possible organic products of the reaction between butan-2-ol and concentrated sulphuric acid at 170°C?

1 Alcohol –OH group forms H-bonds with water molecules.
2 (a) $CH_3CH_2CH_2CH_2OH + [O] \longrightarrow CH_3CH_2CH_2CHO + H_2O$
$CH_3CH_2CH_2CH_2OH + 2[O] \longrightarrow CH_3CH_2CH_2COOH + H_2O$
(b) $CH_3CH_2CHOHCH_3 + [O] \longrightarrow CH_3CH_2COCH_3 + H_2O$
3 $CH_3CH=CHCH_3$ and $CH_3CH_2CH=CH_2$

7.6 Halogenoalkanes

After studying this section you should be able to:

- understand the nature of the polarity in halogenoalkanes
- describe nucleophilic substitution reactions of halogenoalkanes
- explain the relative rates of hydrolysis of different halogenoalkanes
- describe elimination reactions of halogenoalkanes
- understand that reaction conditions can be used to control the products

LEARNING SUMMARY

Halogenoalkanes General formula: $C_nH_{2n+1}X$, where X = F, Cl, Br or I

AQA	M3	OCR	M2
EDEXCEL	M2	SALTERS	M2
NICCEA	M2	WJEC	M2
NUFFIELD	M2

Like alcohols, halogenoalkanes can be primary, secondary or tertiary.

Types and naming of halogenoalkanes

1-bromopropane
primary

2-bromopropane
secondary

2-bromo-2-methylpropane
tertiary

Polarity of halogenoalkanes

Carbon and halogens have different electronegativities and halogenoalkanes have polar molecules with a polar C–X bond.

$C^{\delta+}-F^{\delta-}$
$C^{\delta+}-Cl^{\delta-}$
$C^{\delta+}-Br^{\delta-}$
$C^{\delta+}-I^{\delta-}$

polarity **decreases**

chlorine is more electronegative than carbon – electron flow from carbon to chlorine – dipole produced

The polarity produces an electron-deficient carbon atom, $C^{\delta+}$ which is important in the reactions of halogenoalkanes. The polarity decreases from fluorine to iodine, reflecting the decrease in electronegativity down the halogen group.

Nucleophilic substitution reactions of halogenoalkanes

AQA	M3	OCR	M2
EDEXCEL	M2	SALTERS	M2
NICCEA	M2	WJEC	M2
NUFFIELD	M2

The electron-deficient carbon atom of the polar $C^{\delta+}-X^{\delta-}$ bond attracts nucleophiles, allowing a nucleophilic substitution reaction to take place in which the nucleophile replaces the halogen atom (see also alcohols, page 130).

The hydrolysis of halogenoalkanes

A **hydrolysis** reaction is a reaction in which water breaks a bond.

KEY POINT

Reaction with aqueous hydroxide ions, OH⁻(aq) produces an alcohol.

$$C_2H_5Br + OH^-(aq) \xrightarrow[reflux]{OH^-/H_2O} C_2H_5OH + Br^-$$

The hydroxide ion, OH⁻, behaves as a nucleophile.

Mechanism of nucleophilic substitution (not EDEXCEL or WJEC)

Primary halogenoalkanes react in a one-step mechanism.

NaOH(aq) is used as a source of OH– (aq).

The OH– ion behaves as a nucleophile by donating an electron pair.

Note the usefulness of the curly arrow in showing movement of an electron pair:

- the :OH– nucleophile donates its electron pair to the electron-deficient $C^{\delta+}$ of the bromoalkane, forming a covalent bond;
- the C–Br bond is broken by heterolytic fission with loss of its electron pair to form a :Br– ion.

Comparing the hydrolysis reactions of halogenoalkanes

The rates of hydrolysis of different halogenoalkanes can be investigated as follows:

An alternative method uses water itself for the hydrolysis. The halogenoalkane is heated with a mixture of water, ethanol and $AgNO_3$(aq). The rate of hydrolysis is monitored directly by observing precipitation of replaced halide ions as silver halides.

- Alkaline hydrolysis with $NaOH(aq)/H_2O$, reflux:

$$R\text{-}X + OH^- \longrightarrow R\text{-}OH + X^-$$

- Acidification with dilute nitric acid. This remove excess NaOH which would otherwise form a precipitate with the silver nitrate in the next stage, preventing detection of any silver halides.
- Addition of $AgNO_3$(aq) shows the presence of aqueous halide ions (see also Halogens, page 77).

$$Ag^+(aq) + X^-(aq) \longrightarrow AgX(s)$$

The intensity of any precipitate indicates the extent and rate of the hydrolysis.

$C^{\delta+}\text{-}F^{\delta-}$
$C^{\delta+}\text{-}Cl^{\delta-}$ polarity
$C^{\delta+}\text{-}Br^{\delta-}$ **increases**
$C^{\delta+}\text{-}I^{\delta-}$

bond	bond enthalpy /kJmol^{-1}
C–F	467
C–Cl	364
C–Br	290
C–I	228

> **Two factors – which is more important?**
> - **Polarity** predicts that the more polar C–F bond would attract water most easily and would give the fastest reaction.
> - **Bond enthalpy** predicts that the C–I bond would be broken most easily and would give the fastest reaction.
>
> For this reaction, bond enthalpy is more important than polarity.
> The rate of hydrolysis increases as the C–X bond weakens.

KEY POINT

Further examples of nucleophilic substitution

Nucleophilic substitution of halogenoalkanes is used in organic synthesis to introduce different functional groups onto the carbon skeleton. The reaction scheme below shows how halogenoalkanes react with ammonia and cyanide ions.

For hydrolysis, water is used as the solvent. For other nucleophilic substitutions, it is important that water is absent or hydrolysis will take place.

Ethanol is used as the alternative solvent.

Elimination reactions of halogenoalkanes

AQA	M3	NUFFIELD	M2
EDEXCEL	M2	OCR	M2
NICCEA	M2	WJEC	CH2

When a halogenoalkane is refluxed with hydroxide ions in **anhydrous** conditions (using NaOH in ethanol as a solvent at 78°C), an **elimination** reaction takes place.

$$H-CH_2-CH_2-Br + OH^- \xrightarrow[\text{reflux}]{OH^-/\text{ethanol}} CH_2{=}CH_2 + Br^- + H_2O$$

Mechanism (AQA only)

Water is eliminated from the primary halogenoalkane in a **one-step** mechanism:

$$H-CH_2-CH_2{-}Br \xrightarrow{\text{single step}} CH_2{=}CH_2 + H_2O + {:}Br^-$$

with OH^- attacking.

The hydroxide OH^- ion behaves as a base, accepting a proton H^+ to form H_2O.

Elimination versus hydrolysis

AQA	M3	NUFFIELD	M2
EDEXCEL	M2	OCR	M2
NICCEA	M2	WJEC	CH2

The reaction of hydroxide ions with halogenoalkanes forms different organic products, depending upon the reaction conditions used.

Use different reaction conditions to control the type of reaction.

Commonly tested in exams.

> **KEY POINT**
>
> aqueous conditions $\longrightarrow$ alcohol: OH^- acts as a nucleophile
> anhydrous conditions $\longrightarrow$ alkene: OH^- acts as a base

The ability to control *which* reaction takes place solely by changing the conditions is an important tool for the synthetic chemist.

Uses of halogenoalkanes

Chlorofluorocarbons, CFCs (refrigerants, propellants, blowing polystyrene, dry cleaning, degreasing agents).

Chloroethene and tetrafluroethene to produce the plastics *pvc* and *ptfe*.

Halogenoalkanes as synthetic intermediates in chemistry.

Polychlorinated compounds as herbicides.

ELIMINATION
OH⁻ as a base
Accepts H⁺ to form H₂O

SUBSTITUTION
OH⁻ as a nucleophile
Donates an electron pair

OH⁻/ethanol reflux

OH⁻/H₂O reflux

ELIMINATION
ethanol solvent
anhydrous conditions

SUBSTITUTION
water solvent
hydrated conditions

Progress check

1 What is meant by the term nucleophilic substitution?

2 Why does 1-bromopropane react with nucleophiles but propane does not?

3 Why do bromoalkanes react more readily than chloroalkanes?

4 What are the organic products for the reactions of 1-bromopropane with:
(a) $NaOH/H_2O$; (b) $NH_3/ethanol$; (c) $NaOH/ethanol$?

1 Replacement of an atom or group in an organic molecule by a nucleophile.
2 C–Br bond is polar, C–H bonds are not polar.
3 C–Br bond is weaker than C–Cl bond.
4 (a) $CH_3CH_2CH_2OH$ (b) $CH_3CH_2CH_2NH_2$ (c) $CH_3CH{=}CH_2$.

Sample question and model answer

For open-ended questions of this type, there are usually more marking points than the total – you have some breathing space.

1

(a) Explain, using examples, the following terms used in organic chemistry.

(i) general formula

This is the simplest algebraic representation for any member of a series of organic compounds, e.g. alkanes: C_nH_{2n+2} ✓

(ii) homologous series

This is a series containing compounds with similar chemical properties ✓ with each successive member differing by CH_2 ✓, e.g. 'alkanes' ✓

Do make sure that you choose valid examples. In the actual examination, many candidates managed to confuse homolytic and heterolytic fission – poor preparation!

(iii) homolytic fission and heterolytic fission

Fission means the breaking of a bond. ✓

In homolytic fission, one electron from the bond is released to each atom with free radicals forming. ✓ e.g.: $Cl_2 \longrightarrow 2Cl\bullet$ ✓

In heterolytic fission, both electrons in the bonding pair are released to one of the atoms with ions forming ✓ e.g.: $Br_2 \longrightarrow Br^+ + Br^-$ ✓

9 marks $\longrightarrow$ [7]

Always explain your working. This type of problem is often easier than it looks.

You should look back at the question many times to check that you have answered all parts.

(b) Cracking of the unbranched compound **E**, C_6H_{14}, produced the saturated compound **F** and an unsaturated hydrocarbon **G** (M_r, 42). Compound **F** reacted with bromine in u.v. light to form a monobrominated compound **H** and an acidic gas **I**. Compound **G** reacted with hydrogen bromide to form a mixture of two compounds **J** and **K**.

(i) Use this evidence to suggest the identity of each of compounds **E** to **K**. Include equations for the reactions in your answer.

E is unbranched. ∴ **E** = $CH_3CH_2CH_2CH_2CH_2CH_3$ ✓

G has M_r = 42; must have a double bond: must contain 3 carbons ∴ **G** = C_3H_6. ✓

Notice that everything 'works': questions are designed in this way. If you find yourself with 'strange' structures containing functional groups that you haven't seen before, you have probably made a mistake. This is the time to check back through your logic.

Cracking produces an alkane and an alkene, **F** is an alkane: **F** = C_3H_8 ✓

The equation is: $C_6H_{14} \longrightarrow C_3H_6 + C_3H_8$ ✓

Alkanes react with bromine in u.v. by substitution. ∴ **H** = C_3H_7Br ✓, **I** = HBr ✓

The equation is: $C_3H_8 + Br_2 \longrightarrow C_3H_7Br + HBr$ ✓

Alkenes react with bromine by addition. HBr can be added across a double bond in two ways to give: **J** and **K** as $CH_3CH_2CH_2Br$ ✓ and $CH_3CHBrCH_3$. ✓

The equation is: $C_3H_6 + HBr \longrightarrow C_3H_7Br$ ✓

(ii) Oil companies often 'reform' compounds such as **E**. Explain why this is done and suggest two organic products of the reforming of **E**.

Any 2 examples here formed from C_6H_{14}. These could be branched hydrocarbons, aromatic or alicyclic – but any chosen **must** contain 6 carbons (Read the question!).

Reforming converts low-demand fractions into more useful hydrocarbons ✓ e.g. compound **E**, hexane could be converted into benzene ✓ or 2-methylpentane.

(iii) Predict the structure of the polymer that could be formed from compound **G**.

When drawing a polymer: 'side bonds' are always needed. Always show the repeat unit.

✓✓ (1 for no double bond, 1 for correct skeleton) [14]

Practice examination questions

1 Petrol and camping gas are examples of fuels that contain hydrocarbons.

(a) Petrol is a mixture of alkanes containing 6 to 10 carbon atoms. Some of these alkanes are isomers of one another.

(i) Explain the term *structural isomers*.

(ii) State the molecular formula of an alkane present in petrol.

[2]

(b) The major hydrocarbon in camping gas is butane. Some camping gas was reacted with chlorine to form a mixture of isomers.

(i) What conditions are required for this reaction to take place?

(ii) Two isomers, **A** and **B**, were separated from the mixture. These isomers had a molar mass of 92.5 g mol^{-1}.
Deduce the molecular formula of these two isomers.

(iii) Draw the displayed formulae of **A** and **B** and name each compound.

[6]
[Total: 8]
Cambridge Chemistry Foundation Q3 June 1998 (modified)

2 Two types of isomerism in organic compounds are structural isomerism and *cis-trans* isomerism.

(a) Draw displayed formulae for:

(i) two structural isomers of $C_2H_4Cl_2$;

(ii) the *cis* and the *trans* isomers of $C_2H_2Cl_2$.

[4]

(b) Suggest how the *cis-trans* isomers of $C_2H_2Cl_2$ could be converted into one of the structural isomers of $C_2H_4Cl_2$.

[2]

(c) Under certain conditions, the *cis-trans* isomers of $C_2H_2Cl_2$ can polymerise.

(i) Draw a short section of the polymer formed showing **two** repeat units.

(ii) Polymers are often disposed of by combustion. In this process, in addition to carbon dioxide and steam, toxic gases may be formed. Suggest **two** toxic gases that could be formed during the combustion of the polymer in **c(i)**.

[3]
[Total: 9]
Cambridge Chemistry Foundation Q3 Feb 1997

Practice examination questions *(continued)*

3 (a) Identify compounds **A–C** in the flowchart below.

$$CH_3CH=CHCH_3 \xrightarrow[\text{}]{\text{HBr}} A \xrightarrow[\text{heat}]{\text{OH}^-\text{(aq)}} B \xrightarrow[\text{heat}]{\text{Cr}_2\text{O}_7^{2-}/\text{H}^+} C$$

[3]

(b) Name the type of reaction taking place in each reaction above and state the role of the reagents.

   (i) $CH_3CH=CHCH_3 \longrightarrow A$

   (ii) $A \longrightarrow B$

   (iii) $B \longrightarrow C$

[6]

(c) What reagents and conditions would be needed to convert but-2-ene directly into compound **B**?

[2]

(d) Draw a short section of the addition polymer formed by but-2-ene showing two repeat units.

[2]

[Total: 13]

4 Reduction of a branched alkene **A** with hydrogen formed a compound **B**. On analysis, **B** had the composition by mass of C, 82.8 %; H, 17.2 % ($M_r = 58$). Reaction of **A** with steam in the presence of a catalyst produced a mixture of two isomers **C** and **D**.

Identify possible structures for compounds **A–D**. Your answer should show all working and include structural formulae for each structure and equations for each reaction.

[9]

Cambridge Chemistry Foundation Q7 Feb 1997 (modified)

5 (a) There are four structural isomers of molecular formula C_4H_9Br. The structural formula of two of these isomers are compound **A**: $CH_3CH_2CHBrCH_3$ and compound **B**: $(CH_3)_3CBr$.

   (i)   Draw the remaining two structural isomers.

   (ii) Give the name of **B**.

[3]

(b) All four structural isomers of C_4H_9Br undergo similar reactions with ammonia.

   (i)   What type of reaction is involved in these reactions?

   (ii) Draw the structural formula of the product formed by the reaction of **A** with ammonia.

[2]

(c) Ethene, C_2H_4, reacts with bromine to give 1,2-dibromoethane.

   (i)   Give the name of the mechanism involved.

   (ii) Show the mechanism for this reaction.

[4]

[Total: 9]

AEB Module Paper 6 Q4 Jun 1998 (modified)

Practice examination answers

Chapter 1 Atomic structure

1 (a)

particle	relative mass	relative charge
proton	1	1+	✓
neutron	1	0	✓
electron	1/1840	1–	✓

[3]

(b) (i)   neutron ✓; ^7Li has 1 more neutron ✓

(ii)   electron ✓; $^{32}S^{2-}$ has 2 more electrons ✓

(iii)   proton ✓; $^{40}Ca^{2+}$ has 1 more proton ✓

[6]
[Total: 9]

2 (a) Difference: different numbers of neutrons. ✓

Similarities: same numbers of protons ✓ and electrons. ✓   [3]

(b) ^7Li ✓   [1]

(c) (i) $1s^22s^22p^63s^23p^6$ ✓;   (ii) p block ✓   [2]

(d) (i) $Al(g) \longrightarrow Al^+(g) + e^-$; ✓✓   (ii) Electron lost is in a different shell ✓, closer to the nucleus ✓, less shielding ✓, experiencing a greater nuclear attraction ✓.   [6]
[Total: 12]

3 (a) Energy required to remove 1 electron ✓ from each atom in 1 mole ✓ of gaseous atoms ✓ to form 1 mole of gaseous 1+ ions.   [3]

(b) Nucleus of beryllium has 1 more proton ✓. The extra electron in beryllium is in the same sub-shell (2s)/The attraction of the 2s electrons from the nucleus is greater in Be than in Li ✓.   [2]

(c) $1s^22s^22p^1$ ✓   [1]

(d) The extra electron in boron is in the 2p sub-shell ✓. Less energy is required to remove an electron from the higher energy 2p sub-shell ✓.   [2]

(e) 1200 kJ mol^{-1}. The value will be greater than that of C ✓. The extra electron in nitrogen is in the same sub-shell (2p)/The attraction of the 2p electrons from the nucleus is greater in N than in C ✓.   [2]
[Total: 10]

Chapter 2 Atoms, moles and equations

1 (a) (i)   0.011 ✓;   (ii) 0.055 ✓✓;

(iii)   132 cm^3 using 24 dm^3; 136 cm^3 using $pV=nRT$ (AQA and WJEC) ✓✓✓   [6]

(b) 0.167 M ✓✓✓   [3]
[Total: 9]

2 (a) (i)   510 ✓✓;   (ii) 1020 ✓;   (iii) 204 dm^3 ✓   [4]

(b) 51 kg ✓✓✓   [3]

(c) NaOH is a strong alkali ✓; products of NaOH are soluble ✓   [2]
[Total: 9]

3 (a) (i)   0.075 ✓;   (ii) 3.25 g ✓✓   [3]

(b) 2 mol dm^{-3} ✓✓   [2]

(c) (i)   good conductor of heat/low melting point ✓

(ii)   high reactivity of sodium/molten metals are hot/burn ✓   [2]
[Total: 7]

Chapter 3 Bonding and structure

1 (a) A shared ✓ pair ✓ of electrons [2]

(b) The lone pair of electrons from N of an ammonia molecule ✓ bonds covalently to H⁺ to form an ammonium ion ✓ [2]

(c) 109.5° ✓. 107° ✓. Bonding angle is reduced by repelling effect of a lone pair which is greater than that of a bonded pair ✓ [3]

(d) Hydrogen bonding ✓. This arises from dipole-dipole attraction ✓ between $H^{\delta+}$ of an ammonia molecule and $N^{\delta-}$ on a different ammonia molecule ✓. [3]

[Total: 10]

2 Electron pairs repel one another ✓ and assume a position as **far away** from one another as possible ✓. Lone pairs repel more than bonded pairs ✓. Diagrams are shown below:

(a) [diagram: C with H, bond angle 109.5°] ✓✓

tetrahedral ✓

(b) [diagram: N with H, bond angle 107°] ✓✓✓

pyramidal ✓

(c) [diagram: B with F, bond angle 120°] ✓✓

trigonal planar ✓

(d) [diagram: S with F, bond angle 104.5°] ✓✓✓

non-linear ✓

1 mark for each diagram
1 mark for each bond angle
1 mark for lone pairs on NH_3 and SF_2

[Total: 17]

3 (a) (i) covalent ✓; (ii) van der Waals' forces ✓. [2]

(b) $I_2(g)$ ✓. [1]

(c) Weak van der Waals' forces break on vapourising ✓. [1]

(d) Strong ✓ ionic bonds ✓ hold the lattice together. [2]

(e) They vibrate faster ✓. [1]

(f) On melting, the ionic bonds remain close together and are just loosened from their rigid lattice positions ✓. On vapourising, the ions must be separated, breaking the ionic attraction ✓. [2]

[Total: 9]

4 (a) Water has hydrogen bonding ✓. This is dipole-dipole attraction ✓ between $H^{\delta+}$ of a water molecule and a lone pair ✓ on $O^{\delta-}$ on a different water molecule ✓. [4]

(b) $CHCl_3$ has dipole-dipole interactions ✓ between $H^{\delta+}$ of a $CHCl_3$ molecule ✓ and $Cl^{\delta-}$ on a different $CHCl_3$ molecule ✓. any 2 ⟶ [2]

(c) Ar has van der Waals' forces only ✓. These are caused by oscillating dipoles ✓ which cause induced dipoles on another Ar atom ✓. H-bonding > dipole-dipole > van der Waals' ✓. The intermolecular forces are broken on boiling ✓. The stronger the intermolecular forces, the higher the boiling point ✓. any 4 ⟶ [4]

[Total:10]

Chapter 4 Periodic Table

1 Across Period 3, atomic radius decreases ✓ because there are more protons in the nucleus across the period ✓ increasing the attraction ✓ on the electrons in the same outer shell ✓.

Down Group 1, atomic radius increases ✓ as more shells are being added ✓. The shielding effect increases as the number of inner shells increases ✓. The attraction experienced by the outer electrons from the nucleus decreases ✓. [Total: 8]

2 (a) (i) $Mg(s) + H_2O(g) \longrightarrow MgO(s) + H_2(g)$ ✓

(ii) $2Mg(s) + O_2(g) \longrightarrow 2MgO$ ✓

(iii) $Mg(s) + 2H_2O(l) \longrightarrow Mg(OH)_2 + H_2(g)$ ✓ [3]

(b) (i) On descending outer electrons are further from the nucleus ✓ with greater shielding ✓. Therefore, outer electrons are removed more easily.

(ii) Reaction with O_2/Cl_2/named acid ✓ [3]

(c) (i) very high melting point/strong forces holding lattice together ✓

(ii) CaO: cement ✓; $Mg(OH)_2$: indigestion remedy ✓ *many others possible*. [3]

[Total: 9]

3 (a) On descending the halogens, the atomic radii increase ✓ resulting in less nuclear attraction at the edge of the atom ✓. There are more electron shells between the nucleus and the edge of the atom to shield the nuclear charge ✓. The nuclear attraction at the edge of the atom decreases down the group ✓. Therefore the oxidising power of the halogens decreases down the group ✓. [5]

   (b) (i) $Cl_2 \longrightarrow NaCl$ (ox no: Cl 0 → –1) ✓ reduction ✓;

   $Cl_2 \longrightarrow NaClO_3$ (ox no: Cl 0 → +5) ✓ oxidation ✓

   (ii) 126 g ✓✓✓

   (iii) $Cl_2 + 2NaOH \longrightarrow NaOCl + NaCl + H_2O$ ✓; bleach ✓ [9]

   [Total: 14]

4 (a) (i) Electronegativity is a measure of the attraction of an atom in a molecule for the pair of electrons in a covalent bond ✓.

   (ii) Electronegativity decreases from fluorine to iodine ✓. The atomic radius increases down the group ✓. The shielding effect also increases down the group as the number of electron shells increases ✓. The smaller the halogen atom, the greater the nuclear attraction experienced by the bonding electrons and the greater the electronegativity ✓. [5]

   (b) The boiling points increase from fluorine to iodine ✓. The number of electrons increase ✓, increasing the van der Waals' forces between molecules ✓. [3]

   (c) With aqueous sodium chloride, there is no change ✓. With aqueous sodium iodide, the solution turns dark brown ✓ as iodine is displaced ✓.

   $Br_2(aq) + 2I^-(aq) \longrightarrow 2Br^-(aq) + I_2(aq)$ ✓ [4]

   (d) I⁻ is a stronger reducing agent than Cl⁻ ✓ because I⁻ reduces Br_2, whereas Cl⁻ does not reduce Br_2 ✓. [2]

   [Total: 14]

Chapter 5 Chemical energetics

1 (a) $Al(s) + 1½Cl_2(g) \longrightarrow AlCl_3(s)$ ✓✓ [1, equation; 1, state symbols] [2]

   (b) 100 kPa ✓ [1]

   (c) +269 kJ mol⁻¹ ✓✓✓ [3]

   [Total: 6]

2 (a) (i) The enthalpy change that takes place when one mole of a substance ✓ reacts completely with oxygen ✓ under standard conditions, all reactants and products being in their standard states; (ii) 298K ✓ [3]

   (b) –727 kJ mol⁻¹ ✓✓✓ [3]

   (c) Activation energy must be overcome/energy is needed to break bonds ✓. [1]

   [Total: 7]

3 (a) The enthalpy change that takes place when one mole of a compound ✓ in its standard states is formed from its constituent elements ✓ in their standard states under standard conditions. ✓. [3]

   (b) $C(s) + 2H_2(g) \longrightarrow CH_4(g)$ ✓✓ [1, equation; 1, state symbols] [2]

   (c) –76 kJ mol⁻¹ ✓✓✓ [3]

   (d) C–H: +412 kJ mol⁻¹ ✓; C–C: +348 kJ mol⁻¹ ✓✓ [3]

   [Total: 11]

Chapter 6 Rates and equilibrium

1 (a) Activation energy is the minimum energy required for a reaction to take place ✓. [1]

(b) (i) ✓✓✓; (ii) ✓

(iii) A catalyst provides an alternative reaction route with a lower activation
energy. ✓✓ [6]

(c) (i) ✓✓✓; (ii) ✓✓

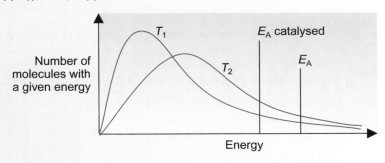

[5]
[Total: 12]

2 (a) (i) A homogeneous equilibrium has all equilibrium species in the same phase ✓.
All species in this equilibrium are gaseous ✓.

(ii) The forward reaction proceeds at the same rate as the reverse reaction ✓.
The concentrations of reactants and products are constant ✓. Equilibrium can
only be achieved in a closed system ✓. [5]

(b) Endothermic ✓. With increased temperature an equilibrium shifts in the endothermic
direction ✓. [2]

(c) (i) Equilibrium position is unaffected ✓. There are equal numbers of gaseous
moles on either side of the equilibrium ✓.

(ii) Rate increases ✓. Increasing the pressure also increases the concentration ✓. [4]
[Total: 11]

Chapter 7 Organic chemistry

1 (a) (i) same **molecular** formula with a different arrangement of atoms/structural formula ✓.

(ii) allow any alkane formula from C_6 to C_{10}, e.g. C_8H_{18} ✓ [2]

(b) (i) u.v. / sunlight ✓; (ii) C_4H_9Cl ✓

(iii) any correct structures ✓✓ with names ✓✓, e.g.

```
    H   H   H   H              H   H   Cl  H
    |   |   |   |              |   |   |   |
H—C—C—C—C—Cl          H—C—C—C—C—H
    |   |   |   |              |   |   |   |
    H   H   H   H              H   H   H   H
```
1-chlorobutane   2-chlorobutane

[6]
[Total: 8]

Organic chemistry (continued)

2 (a)(i) correct structures

✓   ✓

(ii) correct structures

✓   ✓   [4]

(b) H_2 ✓ and a named catalyst/Ni/Pt/Pd ✓.   [2]

(c) (i)

✓

(ii) any two from CO, HCl, Cl_2, $COCl_2$, CCl_4, $CHCl_3$ ✓✓   [3]
                    [Total: 9]

3 (a) A = $CH_3CH_2CHBrCH_3$ ✓;   B = $CH_3CH_2CHOHCH_3$ ✓;   C = $CH_3CH_2COCH_3$ ✓   [3]

(b) (i) addition ✓, electrophile ✓;

(ii) substitution ✓, nucleophile ✓;

(iii) oxidation ✓, oxidising agent ✓.   [6]

(c) steam ✓ and H_3PO_4 ✓   [2]

(d)

no double bond ✓, correct skeleton ✓   [2]
                    [Total: 13]

4 Empirical formula of **B** = C_2H_5 ✓✓; Molecular formula of **B** = C_4H_{10} ✓

A = $(CH_3)_2C=CH_2$ ✓; **B** = $(CH_3)_2CHCH_3$ ✓;

C and **D** = $(CH_3)_2CHCH_2OH$ ✓ and $(CH_3)_2COHCH_3$ ✓;

$C_4H_8 + H_2 \longrightarrow C_4H_{10}$ ✓

$C_4H_8 + H_2O \longrightarrow C_4H_9OH$ ✓

                    [Total: 9]

5 (a) (i) $CH_3CH_2CH_2CH_2Br$ ✓ and $(CH_3)_2CHCH_2Br$ ✓;

(ii) 2-bromo-2-methylpropane ✓   [3]

(b) (i) nucleophilic substitution ✓;   (ii) $CH_3CH_2CHNH_2CH_3$ ✓   [2]

(c) (i) electrophilic addition ✓

(ii) arrows stage 1 ✓;   arrow stage 2 ✓;   carbonium ion ✓

                    [4]
                    [Total: 9]

The Periodic Table

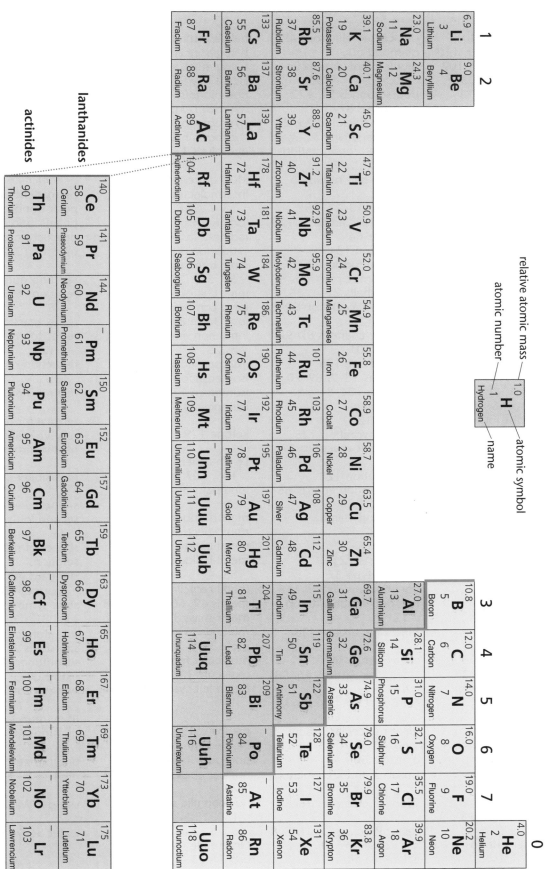

Index

acid-base titrations 40
acids 40, 78, 89, 109, 110
activation energy 99, 102
addition reactions 123
alcohols 116, 127–131
aldehyde 129
alkanes 114–5, 120–1, 132–4
alkenes 115, 122–5
alkyl group 115–16
aluminium extraction 83
ammonia 103, 107–8, 110
amount of substance 33–4
atomic mass 30–2
atomic number 19, 64–5
atomic radius 25, 65, 76
atomic structure 19–27
atoms 19–20, 33
Avogadro constant 33

balancing equations 38–9
bases 40, 109, 110
Basic Oxygen Process 82
beryllium 27, 75
blast furnace 81
bleach 79, 80
boiling points 66–7
Boltzmann distribution 99–100, 101, 102
bond enthalpy 94–5
bond fission 116–17
bonded pair electrons 50–1
bonds 58–61
boron 27
Brønsted-Lowry theory 109

carbon chain extension 133–4
carbon isotopes 20
carbon-12 30, 33, 34
carboxylic acid 129
catalysis 102–3
catalyst 102, 106, 108
catalytic converters 103
chemical bonding 44–9
chlorine 79, 80, 103
cis-trans isomerism 122
combustion 120, 128
covalent bonding 47–8, 52–3, 61
covalent compounds 59–60, 61
cracking 118–19
crude oil products 118–19

dative covalent bonds 48
dehydration 130

delocalised electrons 49
diamond 60
dipole-dipole interactions 55, 56
double bonds 51, 122
electrolysis 82–3
electron shielding 25, 26
electronegativity 52–3, 76
electronic configurations 23, 24, 27, 71
electron-pair repulsion 50–1
electrons 19–24, 25
electrophilic addition 123
elements 64
elimination 134
empirical formula 37
endothermic reactions 88
energy shells 21–4, 26
enthalpy calculation 90–2, 94–5
enthalpy changes 87–92, 94–5, 101
epoxyethane 126
equations 38–9, 69–70
equilibria 104–8
esters 131
ethanol 128
ethene hydration 128
exothermic reactions 87

fermentation 128
first ionisation energy 66
first law of thermodynamics 87
fission 116–17
flame tests 72
formula type 37
fractional distillation 118
free-radical substitution 121
functional groups 115

gas volumes 36
giant ionic lattices 46, 58, 61
giant metallic lattice 60, 61
giant molecular structure 60, 61
graphite 60
Group 1 and 2 elements 71–5
Group 7 elements 76–80

Haber Process 103, 107–8
half-equations 70
halide reactions 78

halide testing 77
halogen reactivity 77
halogenoalkanes 132–4
Hess's law 91–2
homologous series 114
hydrocarbons 113–14
hydrogen bonds 56–7, 61
hydrogen halides 78
hydrolysis 132–4

ideal gas equation 36
intermolecular forces 55–7, 58
iodine 59
iodine-thiosulphate titration 80
ionic bonding 45–7, 52, 53, 54
ionic charge 46
ionic compounds 58–9
ionic formulae 46–7
ionisation energy 25–7, 66
ions 25
iron production 81–2
isomerisation 119
isomerism 114–15, 122
isotopes 20

ketone 129

Le Chatelier's Principle 105, 107
lone pairs 50–1

mass number 20
mass spectrometer 31
metal extraction 81–4
metal halide reduction 83–4
metal recycling 84
metals 49, 60, 61, 67, 68
molar mass 34
molarity 35
mole 33–6
molecular formula 37
molecule 32, 47
multiple covalent bonding 48

neutrons 19, 20
Noble Gases 44, 55
non-polar bonds 53
nuclear charge 25, 26, 65, 76
nucleophilic substitution 130, 132–3

Octet Rule 44, 48
orbitals 21–4, 27

organic formulae 113
organic nomenclature 113, 116, 122, 127, 132
oxidation 68–70, 128
oxidation number 69–70
ozone 103

Pauling scale 52
Periodic Table 24, 26, 46, 52, 64–84, 143
periodicity 65
permanent dipole 53, 56
polar bonds 53
polarisation 54, 127
product quantities 39
protons 19–20

reacting quantities 39
reaction rates 98–101, 102
redox reactions 68–70
reduction 68–70, 78, 124, 129
reduction of metal oxides 81
relative atomic mass 30–1, 34
relative formula mass 32
relative mass 33
relative molecular mass 32, 36
reversible reactions 104–8

s-block elements 71–5
second ionisation energy 26
shapes of molecules 50–1
solutions 35
standard enthalpy changes 88–9
state symbols 38
steel 82
sub-atomic particles 19
sub-shells 21–4, 26–7
substitution 121, 130, 132–3
symmetrical molecules 53

titanium 83–4
titrations 40

unsymmetrical molecules 53

van der Waals' forces 55, 61
water 56–7, 59